# The
# Gardener's
# Almanac

Also by Alan Titchmarsh:

*Non-Fiction*
Trowel and Error
Nobbut a Lad
England, Our England
Knave of Spades
Marigolds, Myrtle and Moles

*Fiction*
Mr MacGregor
The Last Lighthouse Keeper
Animal Instincts
Only Dad
Rosie
Love and Dr Devon
Folly
The Haunting
Bring Me Home
Mr Gandy's Grand Tour
The Scarlet Nightingale
The Gift

# The Gardener's Almanac

## A Treasury of Wisdom and Inspiration through the Year

Written and illustrated by
ALAN TITCHMARSH

HODDER &
STOUGHTON

First published in Great Britain in 2022 by Hodder & Stoughton
An Hachette UK company

I

Copyright © Alan Titchmarsh 2022
Illustrations copyright © Alan Titchmarsh 2022

The right of Alan Titchmarsh to be identified as the
Author of the Work has been asserted by him in accordance
with the Copyright, Designs and Patents Act 1988.

A CIP catalogue record for this title is available from the British Library

Hardback ISBN 978 1 529 38941 8
eBook ISBN 978 1 529 38942 5

Typeset in Imprint by Hewer Text UK Ltd, Edinburgh
Printed and bound in Great Britain by Clays Ltd, Elcograf S.p.A.

Hodder & Stoughton policy is to use papers that are natural, renewable
and recyclable products and made from wood grown in sustainable
forests. The logging and manufacturing processes are expected to
conform to the environmental regulations of the country of origin.

Hodder & Stoughton Ltd
Carmelite House
50 Victoria Embankment
London EC4Y 0DZ

www.hodder.co.uk

*For Hugh from Kew*
*A friend indeed*

# Contents

# Introduction

An almanac, says the dictionary, is 'an annual calendar containing important dates and statistical information such as astronomical data and tide tables'. I suspect much of that information would be of little relevance to a gardener, especially those in cities (who cannot see the stars for a myriad of street lights) and those whose patch of earth is situated some miles from the coast. So while this book does contain some 'statistical information', it is more after the fashion of what used to be called a 'commonplace book' – a seasonal notebook of timely jottings, of quotations and musings, of things that have caught my eye and which give me pleasure as the seasons unfold. Oh, and yes, hints and tips about this and that in the garden as the months go by.

Over the years, I have penned many books that will tell you in detail the 'how', the 'why' and the 'when' of gardening, but this book is rather different. It is what you might call a 'dipper': a book to open at whichever month you're in or else thinking about; to peep into and choose a passage or a fact or quotation that takes your fancy, that inspires you or just makes you feel good. I hope it might surprise you, or make you smile, or raise an eyebrow or a chuckle, or simply make you feel better than you did when you opened up the covers. It will sit happily on your bedside table.

There is certainly a nudge or two – about when to prune your roses, or to encourage you to think about bees and butterflies, or to persuade you to buy a greenhouse or a good tree for winter interest; so it will, I hope, be useful – but above all I want my love of gardens and growing things and the great outdoors to rub off a little, rather than chivvying you into feeling guilty about what you have not done or what you ought to be doing.

Over the centuries many writers have enthused about the joys of the natural world – a world that begins in that patch of earth outside your back door. Their words have delighted and inspired me since I learned to read, and sometimes when we are so busy *doing*, we tend to forget why we bother being out there in the first place.

We garden for so many reasons: because it feeds us body and soul, because it offers sustenance to wildlife as much as to ourselves, but also because, deep down, we feel a duty to our landscape – however minuscule our patch of earth might be – and a need to hand it on to our children and grandchildren in good heart.

We are stewards; we husband the earth – such a lovely but latterly forgotten phrase. Our labours may involve honest toil and a degree of heartache and fatigue, but above all they reward and enrich us for working *with* nature rather than against her.

Oh dear, that sounds rather heavy. Something that this book is not. It is here to remind us all why we tend a patch of earth and the delights that occur as a result of wielding spade and hoe, but also of sitting in the shade in high summer and feeling those welcome rays of sunshine warming our backs on a crisp winter's day.

As almanacs go, this particular one is light on the guilt and heavy, I hope, on the delight. Nothing in life can compare with the pleasure of simply being outdoors and feeling content on a patch of earth we call a garden.

# Sunrise and Sunset in Skipton

I've included the hours of dawn, sunrise, sunset and dusk throughout the year, along with hours of daylight, since every almanac should do that, but I don't expect you to get out of bed at the crack of dawn and grab your spade from the shed. They show how our days lengthen as winter recedes and summer approaches, and how they shorten as we head into autumn and the northern hemisphere tilts further away from the sun. Why Skipton? It is a market town in North Yorkshire. It is where I took my driving test back in the sixties. (I lived in nearby Ilkley and Skipton had the nearest test centre.) It is where certain scenes in my novel *The Gift* were set. I am fond of the place. All of which may be of little consequence to you personally. However, it is also roughly halfway up mainland Britain, which is a more sensible reason for choosing it than mere sentiment.

The times of sunrise and sunset will vary from Skipton as you head north or south. At Wick, on the north coast of Scotland, I understand that there is enough light to play a round of golf at midnight on the longest day, by which time the folk of Eastbourne, in

Sussex by the sea, will have long repaired to the bar or retired with their mug of Ovaltine. Conversely, on the shortest day of the year, 21 December, there are only 6 hours 20 minutes of daylight in Wick and the folk of Eastbourne (well wrapped up) will still be partying on the beach for another hour and a half. This variation in latitude between north and south is why the British Isles, sitting on the shoulder of the globe, are capable of growing sub-arctic plants in Caithness and Sutherland, and Mediterranean flowers in Cornwall and the Isles of Scilly. I rather like that . . .

January

January brings the snow;
Makes your feet and fingers glow.

FLANDERS AND SWAN, *A Song of the Weather*, 1963

Gardeners always think that this year will be better than last. Farmers always know it will be worse. And that, in a nutshell, is the difference between gardeners and farmers – gardeners are the eternal optimists. We see January as a chance to make a fresh start. The snow-drops are pushing up and winter has agreed to think about calling it a day (though it would be foolhardy to imagine that spring will arrive with February.) At least in January the gardener can move rather faster than nature – faster even than the weeds, which, come April, will seem to be on steroids and the feeble gardener is struggling to keep up. (Not that I spend much time weeding in my garden; I plant things too closely to allow room for interlopers. It seems to me to be prefer-able to great swathes of grey earth that lie in wait to offer hospitality to weed seeds . . .)

## Weather

January is a capricious month: it may be 'unseasonably mild', as the forecasters call it, with higher temperatures than average and heavy rainfall that leads to devastating floods. Strong winds may bring down trees. Temperatures may suddenly plummet and the landscape will be locked in winter's icy grip, with snowfall that isolates upland villages for days on end (it used to be weeks, your grandfather will tell you). In short, in January the weather could do anything, and in spite of the fact that news programmes will shriek about gales and storms, January has seen it all before. The days are short, the nights are long, but at least the garden is pretty much asleep and there is time to catch up, to plan, to dream, to redesign . . .

# Day length (Skipton)

|  | 1 January | 31 January |
| --- | --- | --- |
| **Dawn** | 07.43 | 07.17 |
| **Sunrise** | 08.24 | 07.56 |
| **Sunset** | 15.58 | 16.46 |
| **Dusk** | 16.39 | 17.25 |
| **Daylight hours** | 7 hours 28 minutes | 8 hours 48 minutes |

# Feast days for gardeners

| | |
|---|---|
| **1 January** | **New Year's Day** – raise a glass to a new season! |
| **5 January** | **Burns Night** (For gardeners? Well, someone has to grow the neeps and tatties . . .) |
| **5 January** (if you want to be strict) or **6 January (Epiphany)** | **Twelfth Night** – the end of Christmas festivities by which time all Christmas decorations should have come down. |
| **First Monday after Epiphany: 6 January** | **Plough Monday** – the traditional start of the agricultural year when ploughmen returned to the fields. |
| **Last weekend of the month** | **Big Garden Birdwatch** – organised by the Royal Society for the Protection of Birds. |

# Tree of the month

## Silver birch (*Betula pendula*)

Our native silver birch is also rather delightfully known as the lady of the woods, no doubt on account of its feminine elegance. It makes a light-canopied tree, which means it is a good choice for smaller gardens and where space is limited. In the Ukraine, the month of March is known as Berezen – the time of the birches – for that is when the buds begin to burst, but the tree itself is so valued for its bark in winter that I include it in January. Plant the native form in copses and woodland, but for the garden choose *Betula utilis* var. *jacquemontii*, whose bark is much whiter and a great feature in the winter garden when the branches are bare of leaves. Multi-stemmed specimens give you more bark for your bucks. I wash mine each winter to remove green algae so that they positively glow in the winter sunshine. I ignore the strange looks from passers-by.

Height and spread after 20 years: 8 x 4m.

# Music to listen to . . .

Try Hamish McCunn's majestic *Land of the Mountain and the Flood*, unless the weather is such at the time of listening that the title seems a little insensitive. It was written as a paean of praise to Scotland in 1887 and inspired by Sir Walter Scott's *The Lay of the Last Minstrel*:

Land of the heath and shaggy wood,
Land of the mountain and the flood.

George Bernard Shaw slated it after its first performance at Crystal Palace, but a recording in the 1960s led to a resurgence in its popularity. For me it simply shrieks of the

Highlands, rather after the fashion of Mendelssohn's *Hebrides Overture*, which includes the melody known as 'Fingal's Cave' – and Max Bruch's *Scottish Fantasy*. All three are good January listening when you are curled up in front of a fire feeling snug and smug.

## Vegetables from the garden this month

Jerusalem artichokes, sprouting broccoli, Brussels sprouts, winter cauliflower, celeriac, celery, kale, leeks, parsnips, savoy cabbage, spinach, swedes, turnips.

## Vegetables in store

Jerusalem artichokes, beetroot, carrots, onions, parsnips, potatoes, shallots, swedes, turnips.

## Something to muse upon . . .

I find *The Shepherd's Calendar* by John Clare a constant solace. It is a reminder of a time, 200 years ago, when although the country was in the middle of the Industrial Revolution and many deserted the countryside for work in towns and cities, those who still worked on the land felt the barometer of the seasons rather more keenly than we do today.

Withering and keen the Winter comes,
While Comfort flies to close-shut rooms,
And sees the snow in feathers pass
Winnowing by the window-glass;
Whilst unfelt tempests howl and beat
Above his head in chimney seat.
Now, musing o'er the changing scene,
Farmers behind the tavern-screen
Collect; – with elbow idly press'd
On hob, reclines the corner's guest,
Reading the news, to mark again
The bankrupt lists, or price of grain;
Or old Moore's annual prophecies
Of flooded fields and clouded skies;
Whose Almanac's thumb'd pages swarm
With frost and snow, and many a storm,
And wisdom, gossip'd from the stars,
Of politics and bloody wars.
He shakes his head and still proceeds,
Nor doubts the truth of what he reads:
All wonders are with faith supplied, –
Bible, at once, or weather guide.
Puffing the while on red-tipt pipe;
He dreams o'er troubles nearly ripe;
Yet, not quite lost in profit's way,
He'll turn to next year's harvest-day,
And, Winter's leisure to regale,
Hope better times, and – sip his ale.

JOHN CLARE, *The Shepherd's Calendar* (January), 1827

# Flower of the month

## Snowdrop

No other flower is so eagerly awaited or so heartily welcomed after the bleak days of autumn and the first half of winter. There are single snowdrops (*Galanthus nivalis*) and double snowdrops (*G. nivalis* 'Flore Pleno'), which look like the dresses they used to wear on *Come Dancing* in the days before they added 'Strictly'. Layers of tulle . . . But beware; snowdrop addiction is catching. 'Galanthomania' they call it – the affliction of enthusiasts who prize any kind of variation in flower form and pattern, and there are literally hundreds of minor differences. If I were you, I'd give in and enjoy it and become a 'Galanthophile'. I have.

The most expensive single bulb – of the variety 'Golden Fleece' – was sold in 2015 for £1390 plus £4 postage. Other varieties are much more reasonably priced – you can buy 100 ordinary snowdrop bulbs for under £20 – but all are best planted 'in the green' as growing plants rather than as dry bulbs, which may be completely desiccated. They love best any decent well-drained soil in dappled shade among trees and shrubs, though they will also grow in grass if you allow the leaves to remain for six weeks after flowering. I particularly love 'Godfrey Owen', which has a double row of outer tepals (snowdrops don't have 'petals') and varieties like 'Primrose Warburg', whose markings are yellow rather than green and consequently an eye-catching variation.

# Flowers in the garden

*Acacia dealbata* (mimosa), *Chimonanthus fragrans* (wintersweet), *Clematis cirrhosa*, Corylus (hazel), *Crocus tomasinianus*, *Cyclamen coum*, Daphne, *Eranthis hyemalis* (winter aconite), Erica (winter-flowering heathers), Galanthus (snowdrops), *Garrya elliptica*, Hamamelis (witch hazel), Helleborus (hellebores), *Iris reticulata*, *Iris unguicularis*, *Jasminum nudiflorum* (winter jasmine), *Lonicera fragrantissima* (winter-flowering honeysuckle), Mahonia, *Prunus* x *subhirtella* 'Autumnalis', Pulmonaria (lungwort), Sarcococca (Christmas box), *Viburnum bodnantense*, *Viburnum burkwoodii*, *Viburnum farreri* (*V. fragrans*), *Viburnum tinus*..

# Herbs from the garden

Bay, rosemary, thyme. And, if you are lucky, chives and sage.

# A garden to visit in January

Many gardens offer interest if you know where to look, but dodgy weather means it makes sense to find one with some undercover space at this time of year. Treat yourself to a trip to the Eden Project in Cornwall where, under cover of the spectacular biomes, you can wander in a veritable jungle of exotic plants. The coast is not too far away for a brisk stroll by the sea and spring comes earlier to Cornwall than most places. Further north, January is a fine month to enjoy the layout of Studley Royal in Yorkshire and its sophisticated series of water gardens, which sparkle in the winter sunshine. The end of your walk along the banks of the River Skell will terminate at the ruins of Fountains Abbey. Magical!

# Ordering seeds

You know how it is on those dark evenings when you've had a pleasant supper, a glass or two of your favourite tipple and there's nothing on the box? You pick up your mobile or your tablet and you start surfing the net . . . filling a basket. Many a retailer has cause to be grateful for late-night shopping that happens when the store is closed. Clothing and gadgets, furniture and designer handbags can rack up thousands of pounds in a matter of seconds. How much nicer to surf the net and order flower and vegetable seeds. They are less expensive, allow you to dream of a flower- and food-filled summer, and they give you a warm glow when they arrive in a fat envelope in advance of the growing season. They offer a promise of spring to come, and the sooner you order, the less likely you are to be disappointed.

Order your old faithfuls, by all means – it's comforting to know what you're getting – but try something new as well, with maybe a small propagator to add a bit of heat and get started earlier. Then there are seed potatoes and onion sets, begonia seed that is, ounce for ounce, more expensive than gold . . . but I'm getting carried away. And so will you, with any luck . . .

# Bird of the month

## Jackdaw (*Corvus monedula*)

The jackdaw is a member of the crow family in evidence daily at my bird table here in Hampshire – five or six of them at a time – especially in winter, so it comes as a surprise to discover that they are rare in Central London. The jackdaw is a born scavenger, known for making nests in holes in cliffs, churches and cathedrals and in chimney pots, blocking up the flue. If your chimney smokes when you come to light the first fire of autumn, you may have given home to a jackdaw.

In the sixteenth century, they were positively reviled for their depredations in fields of grain, but you will also read stories about how they were kept as pets. Their distinctive 'yap' as they quarrel, and their grey 'hood' along with their 'small malicious serpent-like grey eyes', as the naturalist W.H. Hudson described them, make them easy to pick out among a flock of rooks and crows. They feed on grain, worms, eggs, small mammals and, alas, nestlings of other birds. Their nests of sticks – clumsily placed – can be huge, and they lay one clutch of four to six eggs, pale blue, spotted with brown, which hatch in April or May. It will take the chicks a month to fledge.

There are around 1½ million pairs in the UK and they are found all over mainland Britain except in the far north-west of Scotland and certain of the Hebridean islands.

# Helping nature

There is no doubt in my mind that a garden where nature is kept firmly in check is a garden where sterility reigns. Anyone who has the remotest feel for the natural world wants to create a garden where wildlife is as much at home as we are. Easy to say, eh? More of a challenge when the wildlife in question is a mole throwing up mounds of earth on the lawn, a rabbit gnawing at the bark of a young sapling, or grey squirrels robbing garden birds of food and stripping the bark from young trees. But we battle on in the face of such predations and console ourselves with the song of the robin as we dig alongside him, the chatter of sparrows in the bay tree, and the eve-tide aria of the blackbird from the chimney-tops.

Feeding garden birds is a common occurrence, but more of us should put up nestboxes to encourage them to breed. Those with small holes will suit the tit family; open-fronted boxes are beloved of robins and fly-catchers. Try to site them where they are unlikely to be reached by marauding cats and facing somewhere between north and east so that they are sheltered from prevailing winds and rain that tend to come from the south-west. Clean them out at the end of every nesting season (in autumn) so that birds can use them to roost through the winter. Now is a great time to put them up, since birds are already prospecting to find a decent nursery for their young. You'll have hours of fun watching them nest building and, later in spring, encouraging their young to fly the nest.

## Fruits in store

Late-ripening eating and cooking apples, late-ripening pears, frozen raspberries, and all those soft fruits you popped into the freezer. Jams and marmalade, too.

## Fruit of the month

### Pear

The pear is the most aggravating of fruits, mainly because it ripens from the inside outwards and does so seemingly overnight when you are not looking. The trick is to store them in a cool, dark, frost-free place after harvesting in autumn and to check them . . . well, daily. Sorry. But when they are ripe . . . oh my goodness!

That said, I've never been able to work up much enthusiasm for the slender-contoured 'Conference', which is usually too hard and crunchy to be enjoyable (the grittiness you feel on your teeth is caused by tiny stone-like particles called 'sclereids'), but 'Doyenne du Comice' is a princess among fruits, with juice that will dribble from your chin. The very top of the fruit – the neck – will soften first and give you a clue that the fruit is almost ripe. Grow it in preference to any other variety and harvest it in October to eat in November and December and – if you are lucky – the beginning of January.

# In the kitchen . . .

Carrots are vegetables that tend to be served simply boiled. Those who demand luxury will drop a lump of butter on to them. As well as buttering them, add chopped basil; it really perks them up and the combination of the two flavours is perfect.

# Something to read . . .

We are blessed with countless gardening books that tell us what to do and when, what to grow and how to grow it: books of erudition and experience that set down hard-won information and expertise to save the rest of us from failure. Reference books, we call them, and every gardener will have well-thumbed favourites. But there are also books written by individuals who inspire by the engaging qualities of their prose, who ruminate in a personable way on the experience of growing plants and making gardens. Hugh Johnson is just such a writer. Have a look at his collection entitled *Sitting In The Shade*. It comprises short essays that are perfect for bedtime reading. Do not be put off by the fact that one Alan Titchmarsh wrote the foreword . . .

# Famous gardener of the month

## Gertrude Jekyll (1843–1932)

There's a wonderful painting of Gertrude Jekyll by William Nicholson showing her in profile, sitting in a chair with steepled fingers, a pair of wire-rimmed specs resting on the end of her nose, her grey hair fastened back in a bun. She looks indomitable and rather intimidating, which she undoubtedly was. Jekyll (pronounced to rhyme with treacle) was the high priestess of *Colour Schemes for the Flower Garden* (the title of one of her many books) and her plant combinations really are worth emulating today. While her garden layouts tend to be formal, the planting within her beds and borders is looser and more lyrical.

She often worked with the architect Edwin Lutyens, designing gardens around his houses, of which many were in Surrey. Jekyll herself lived at Munstead Wood near Godalming. In Hampshire, Rosamund Wallinger at The Manor House, Upton Grey, found Jekyll's plans for the garden there and has recreated the original design. It is worth a visit in summer. Stern as she looks in Nicholson's painting, I find myself smiling at the nickname given to her by Lutyens, who was twenty-six years her junior. He called her 'Aunt Bumps'.

# The toolshed

'Oh to be in the toolshed, now that January's here.' There is no finer escape than a shed of your own, with a chair in the corner, a stove if there's room, hooks for tools, boxes for seeds, a radio on a shelf and . . . you see, that's the joy of it . . . a toolshed can be your own personal sanctuary, designed to suit your own particular requirements: practical, yes, but also a place to find solace, silence and solitude, or to enjoy the plaintive notes of Classic FM.

Do not assume that your shed needs to be large; the smaller it is, the cosier it will be and the easier to keep warm. If you have electricity you can work there of an evening doing such valuable jobs as putting your seed packets in alphabetical order, opening a bottle of wine or beer, and oiling your tools. A bucket filled with sharp sand and mixed with a can of sump oil can be left by the door to act as a cleaner for your spade and fork, hoe and trowel – push them into it a few times and the effect is like that of a pan-scrub. Then sit down and listen to the music, or the birdsong . . . you really want one now, don't you?

# Wild flower of the month

## Old man's beard (*Clematis vitalba*)

Look at the way it garlands country hedgerows in winter with its clouds of silky seedheads. Usually an indicator of chalky soil, old man's beard has white summer flowers that seldom get a second glance among the foliage of the hawthorn and blackthorn, field maple and dog roses that support it, but when the leaves of its supportive hosts fall, it comes into all its feathery glory. Its leaves are the food plant of several moths: the Small Waved Umber and Small Emerald moth among them. Its strong vines have been used to make rope and baskets.

*Other local common names*: Traveller's Joy, Father Christmas, Grandfather's Whiskers, Bellywind, Blind-man's Buff, Tuzzy-muzzy, Hedge Feathers, Snow in Harvest, Withywine, Bullbine, Boy's Bacca, Old Man's Woozard, Smoking Cane, Shepherd's Delight, Smokewood.

# Don't forget the bark . . .

With few leaves to admire at this time of year, except those of evergreens, and the dry and coppery ones that cling to beech hedges (a phenomenon known as 'marcescence'), planting trees and shrubs that have bright bark really does brighten up the garden. Look out for the Japanese maple *Acer palmatum* 'Bi Ho', which has brilliantly coloured stems of orange and pink, and the variety 'Aoyagi' that has green stems, which are surprisingly attractive. Birch trees (see above) with their white or cream bark sing out on the dullest day, the paperbark maple *A. griseum* has curls of peeling mahogany on its trunk and branches, and the ornamental cherry *Prunus serrula* has coppery bark that is even shinier.

If you have no room for a tree, plant dogwoods such as *Cornus sanguinea* 'Midwinter Fire' and 'Anny's Winter Orange', which positively glow. Cut them hard back at the end of February to encourage new growth that will be bright next winter. Honestly, they'll take your breath away.

# Things you can do

— The veg patch will need cultivating – digging or forking over – but take your time. There's no rush. You can spread well-rotted manure or garden compost over the surface afterwards. Fork it in a couple of weeks before sowing seeds in March/April depending on the weather.
— Prune your roses and your fruit trees.
— Check fruit and veg in store and throw out any that are rotting.
— Order seeds online or from catalogues – you'll order far too many, but hey . . .
— Trim the old leaves from hellebores so you can see the flowers emerge.
— Sprout seed potatoes by the window in a spare bedroom, shed or garage.
— Cut back any remaining perennials, unless you think they are beautiful or they offer sustenance to wildlife.
— Plant bare-root trees and shrubs. Those that are container grown can be planted all year round.
— Mow the lawn in dry weather if its fluffiness offends you, but set the blades higher than in summer.
— Force rhubarb under a terracotta forcer filled with straw.
— Feed the birds and provide fresh water daily.

# Things you should not do

— Don't walk on frosted grass.
— Don't sow seeds outdoors – it's too cold and too wet.
— Keep off soggy or frozen soil.
— Don't dig for more than twenty minutes at a time or you'll do your back in.
— Avoid being too tidy in hedge bottoms where insects lurk and provide food for birds.

February

February is a vile little month. You're broke, fat, cold and bored.

KIT HESKETH-HARVEY, writing in *Country Life*

Though I am not quite so hard on February as Kit Hesketh-Harvey, it is that moment in the year when spring is closest and yet seems an age in coming: 'Darkest the dawn when day is nigh.' But at least it is the shortest month in the year at 28 days (29 in a leap year – every fourth year) to keep us in line with the solar year. The earliest snowdrops that were opening in January are joined by their friends. February is definitely snowdrop month and the bravest daffodils like 'February Gold' might well live up to their name if the weather is kind. Hazel catkins – those sulphur-yellow lamb's tails – shed their pollen in clouds on the passing breeze.

The catkins themselves are the showy male flowers. Look at other unopened buds along the twigs. Some will erupt with a tiny tuft of crimson stigmas; these are the female flowers (less conspicuous but vital for the production of hazelnuts). Vivid green bluebell leaves are pushing up in woodland (the flowers will be along in a couple of months). In short, if we can get over the depression of a winter that seems never-ending, these are the little nuggets of hope that we latch onto. When the first primrose opens this month, all's right with the world.

# Weather

Although we are well past the shortest day (21st December), February can be absolutely bitter and snowfall is, to put it diplomatically, 'more than moderately likely'. For goodness sake don't let a February mild spell fool you into thinking that winter is over. Keep your seeds firmly in their packets, whatever it says on the back. Unless you have a greenhouse, of course; then you can spend a ridiculous portion of your income on heating and get started 'under glass' with a propagator. (Until you have one, you will be unaware of the raw excitement that such a piece of equipment can engender.)

Wet or frozen ground outdoors will remain too inhospitable for seed sowing, though bare-root hedging plants, shrubs and trees can be planted when the soil is workable. You are also pretty certain to avoid watering them in – unless it is ridiculously dry for weeks on end, which is not terribly likely in February.

# Day length (Skipton)

|  | 1 February | 28 February |
| --- | --- | --- |
| **Dawn** | 07.15 | 06.22 |
| **Sunrise** | 07.52 | 06.56 |
| **Sunset** | 16.50 | 17.44 |
| **Dusk** | 17.27 | 18.14 |
| **Daylight hours** | 8 hours 52 minutes | 10 hours 42 minutes |

# Feast days for gardeners

**(46 days before Easter Sunday)** **Lent** – traditionally a time of privation in the Christian calendar, may start as early as 4 February on the day known as Ash Wednesday.

**14 February** **St Valentine's Day** – traditionally a time to give your beloved red roses. Be different: offer seasonal delights – a bunch of spring flowers: daffodils, tulips and fragrant freesias, or a pot of fragrant daphne; so much nicer, far less expensive and more environmentally friendly than imported roses.

# Tree of the month

## Witch hazel

The witch hazel, *Hamamelis* x *intermedia*, has aspirations to be a tree, but in most gardens it's a large shuttlecock-shaped shrub. The fact that it flowers on bare branches in winter makes it rather an exciting plant to grow. The flowers are large spiders consisting of a cluster of narrow, strap-shaped petals that may be rich crimson – 'Livia', bright amber – 'Jelena', or yellow, in what I think is the best variety – 'Pallida', quite simply because on a dingy, grey day, when the darker varieties merge into the mist, it is more readily visible.

The flowers are scented of citrus and if the weather is too cold to detect the fragrance, exhale on the blooms through your mouth and then inhale through your nose. Unless you have been heavy on the garlic the night before, the warmth of your breath will release the fragrance. The rather curious thing about witch hazels is that the autumn colour of their fading foliage mirrors the colour of their winter flowers – crimson in 'Livia', orange in 'Jelena' and bright yellow in 'Pallida'. Neat, eh? It will grow on all but the chalkiest and shallowest of soils.

Height and spread after 20 years: 3–4m.

# Music to listen to . . .

Well, the obvious one is *Winter* from *The Four Seasons* by Vivaldi, but that's because it is so fitting for this time of year – those staccato notes on the violin that are chillingly discordant always make me think of dripping icicles. You could, of course, listen to *Spring* from the concerti, but I am a confirmed believer in the postponement of gratification. Save that for March. If you need music to suit your bruised soul, seek out *Light for the World* by the Poor Clares of Arundel – the wonderfully angelic voices of the nuns soar to the rafters – or else *Spem in Alium* by Thomas Tallis. Both are extraordinarily ethereal and will lift you to a higher place on a dreary winter's day.

# Digging . . .

Do you have to? Time was when 'double digging' – cultivating the soil on your veg patch to the depth of two spade blades – was an annual winter task. It probably resulted in more backache than you could shake a fork at. I have never double dug since I was asked to do so for a practical examination when I was fifteen. Some gardeners employ a 'no dig' policy on their veg patches – simply tickling over the surface of the soil with a fork in early spring before sowing their crops. I occupy a halfway house. I spread well-rotted garden compost on

my veg patch in winter and fork it into the surface in
spring. Just before sowing my seeds, I level the earth
with a rake and that's it. I have raised beds filled with
good topsoil, and my parsnips are a foot or more long.
No double digging for me . . .

## Vegetables from the garden this month

Jerusalem artichokes, sprouting broccoli, Brussels
sprouts, late cauliflowers, celeriac, celery, kale, leeks,
parsnips, savoy cabbage.

## Vegetables in store

Jerusalem artichokes, beetroot, carrots, onions, parsnips,
potatoes, shallots, swedes, turnips.

## Something to muse upon . . .

Edward Augustus Bowles gardened at Enfield in
Middlesex. A renowned plantsman and garden writer,
his *Handbook of Crocus and Colchicum*, first published
in 1924, became a standard work. He has a crocus
named after him, along with a snowdrop and 'Bowles's
Golden Grass' – *Milium effusum* 'Aureum', which he
brought to public attention. His three slightly

old-fashioned but eminently readable volumes: *My Garden in Spring, My Garden in Summer* and *My Garden in Autumn and Winter* are still worth dipping into. The words 'climate change' and 'global warming' were a century away when he wrote this:

> If one could but arrange that Winter should arrive with snow and ice, say a week before Christmas, and remain with us for three or even more weeks, and then go away, taking the cold winds and snow to the North Pole, allowing the temperature to increase in warmth gradually from day to day until Spring was due, very little harm would befall the treasures of our gardens. Nowadays Winter has such tricky ways that one never knows how to be even with him . . .
>
> There is a wonderful beauty in the old dead stem, the kecksies, of many plants in a soft winter evening's glow. A ruddy light on brown Grasses or Eryngium stems, or against bare Lime twigs, is a source of great pleasure when it is just too dark to see to weed or dig among the small plants. It is a delight to watch the glow fade off the tall Scots Pines, and to listen to the blackbirds chasing one another about among the evergreen shrubs, and noisily claiming their right to certain desirable roosting places. Then when they have settled down, the robins have a few words to say in their harsh clicking winter voices. It may be only kindly good-night to their friends, but it always has rather an angry sound, as though it referred to the encroachment of neighbours on their roosting rights.

Then comes the stillness of a winter evening. A dim grey vision of one of the Winter-moths flits by, and it is time to move towards the house and the tea-table as the tawny owls wake up and call for physic or start hooting. Then again on a frosty morning, every stem, weed, blade of grass or cobweb has its edging of pearls or diamond dust, and I am always glad if I have not yet tidied away the dead stems that look so lovely in their coats of hoar-frost.

E.A. BOWLES, *My Garden in Autumn and Winter*, 1915

# Flower of the month
## Narcissus

Thought to have been named some time in the sixteenth century after the vain youth who admired his reflection in the water – those planted beside streams nod forwards to admire their own beauty. The large trumpet kinds we call daffodils; those with a smaller 'corona' – the correct name for the trumpet – we call by their botanical name. The earliest varieties will flower this month – 'February Gold' is one such – and others will follow well into April. At one time we all grew those 60-cm high 'King Alfred' and 'Carlton' varieties – what a friend of mine refers to as 'cooking daffs' – comparing them to robust cooking apples. Better for most gardens are the miniatures, which do not get bowed down by snow or rain so that the stems break and the flowers are eaten by slugs.

The easiest to obtain is the variety 'Tête-à-Tête', 20cm high, whose bright yellow flowers are sometimes born in pairs atop their stalks. It is wonderfully reliable. 'Jet Fire' has orange trumpets, 'Bell Song' has sulphur yellow trumpets that age to pale salmon (and which flower under my trees of *Prunus* 'Shirotae' at just the right time, and in just the right shade). The latest narcissi will flower in May – 'Old Pheasant Eye' is a fragrant favourite for planting in the long grass of orchards. Allowed only one narcissus in my garden, I would plump for 'Lemon Silk', a 30-cm high beauty of great elegance and reliability. Remember to let the leaves die down 'untrussed' for six weeks after flowering to allow for next year's flowers to be nourished within the bulbs. No knots, and no elastic bands please!

# Flowers in the garden

*Acacia dealbata* (mimosa), Bergenia (elephant's ears), Camellia varieties, Chaenomeles (Japanese quince or cydonia), Chionodoxa, *Clematis cirrhosa*, *Cornus mas* (cornelian cherry), Corylus (hazel), Crocus varieties, *Cyclamen coum*, Daphne, Eranthis (winter aconite), Erica (winter-flowering heathers), *Garrya elliptica*, Hamamelis (witch hazel), Helleborus (hellebores), *Iris histrioides* varieties, *Iris reticulata* varieties, *Iris unguicularis*, Jasminum (winter-flowering jasmine), *Lonicera fragrantissima* (winter-flowering honeysuckle), *Magnolia campbellii*, Mahonia, Narcissus varieties, *Prunus* x *subhirtella* 'Autumnalis' (winter-flowering cherry), Pulmonaria (lungwort), Sarcococca (Christmas box), Scilla (squill), *Viburnum burkwoodii*, *Viburnum farreri* (*V. fragrans*), Viola varieties.

# Herbs from the garden

Bay, rosemary, thyme. New chives may be emerging. Basil in pots on an indoor windowsill.

# A garden to visit in February

This is snowdrop month and some gardens are richly endowed. Hever Castle in Kent is one of them and talks are arranged from time to time. Waddesdon Manor in Buckinghamshire, built by Baron Ferdinand de Rothschild, is an amazing French chateau of a house with formal gardens and thousands upon thousands of snowdrops in its more natural landscape. Londoners can find them on a smaller scale at Chelsea Physic Garden and in great swathes at Kew Gardens.

Check out Welford Park in Berkshire, and Snowdrop Valley at Wheddon Cross on Exmoor. By the River Tweed at Abbotsford in the Scottish borders there are sheets of them, and Colesbourne Park in Gloucestershire was one of the first gardens to open for its plantation of snowdrops alone. In Kent, the garden that belonged to Jane Austen's brother Edward

– Goodnestone Park – opens daily during the snowdrop season. They look wonderful at Howick Hall in Northumberland and Goldsborough Hall in Yorkshire. In short, there will be sheets of snowdrops in a garden or woodland near you. Just go out and find them.

# Undercover gardening

If every gardener needs a shed, then every keen gardener needs a greenhouse, not least because it means we can potter away to our heart's content whatever the weather. What's more, the growing season is extended at either end, with sowings being made earlier and harvests prolonged in summer and autumn by higher temperatures and protection from extremes of weather. But what sort of greenhouse? And what size? What material should it be made of? What will it cost?

Right, let's start with size. Make it as big as possible. It is a truth universally acknowledged that a single green-house (sorry, Miss Austen) is never big enough. Whatever its size, you will always fill it. If you have a sunny wall, a lean-to greenhouse will suit you well, especially if you can access it through a doorway and avoid getting your feet wet. In the garden, make sure a free-standing greenhouse is not overshadowed by trees. It does not really matter whether it runs north/south or east/west as long as it is not shaded for more than an hour or two each day.

Ventilation – in the ridge *and* the sides – is vitally important to prevent it heating up to oven-like proportions in summer. If it can be wired for electricity, you can install lighting and a thermostatically controlled heater to keep tender plants free of frost in winter. Rainwater can be collected in butts or tanks connected to downpipes from the roof. Erect the greenhouse on good foundations – flagstones or a concrete pad.

When it comes to materials, you get what you pay for. Aluminium alloy greenhouses are among the most inexpensive and last well; those made of western red cedar are the most durable timber kinds and will last many years before you need to repair or replace them. The best kind of all, I reckon, is a greenhouse made from powder-coated alloy. It may be white, cream, pale green (or any other tasteful colour of your choice from the Farrow and Ball range) and will be virtually maintenance free, apart from an annual washing down. The price may make your eyes water, but your greenhouse will be a thing of beauty and a joy forever.

How can you possibly resist the lure of raising your own plants, growing your own tomatoes, cucumbers and melons, and filling it with flowers in summer or winter to gladden your heart?

# Bird of the month

## Blue tit (*Parus caeruleus*)

Originally known as the Tom tit or titmouse, the blue tit is the most acrobatic of our garden birds and the most frequent visitor to hanging bird feeders. Once upon a time, when full cream milk was delivered to our doorstep in foil-topped bottles, the blue-tits would peck open the tops to drink the cream. The first record of this occurrence is in Southampton in 1921. The habit spread until, by the 1940s, it was widespread. They must bemoan the reduction in deliveries and the popularity of semi-skimmed milk. Take a close look at their glorious plumage – yellow, blue, green, grey and white, with that blue crown and black stripe across the eye. The song is a rapid 'tsee-tsee-tsee' or a 'churr'.

This month, blue tits will be prospecting for hollows in trees and nest boxes where they will build a nest of moss and any available fibres. The females will lay between seven and twelve white eggs speckled with red-brown that will hatch in April/May. The chicks will fledge in around three weeks. Food is seeds, nuts, insects and small grubs. There are around 3½ million pairs across the UK.

# Topiary

You don't *have* to fill your garden with peacocks sculpted from holly, yew and box, though you'll make passers-by smile if you do. Over the years I've crafted a few peacocks from yew, but my tastes now are more restrained (perhaps that comes with age and the time needed to see them fully fledged). I have never regretted having cones and orbs and lollipops of yew and box in my garden. They bring shape, form, structure and texture to an otherwise flat plot in winter and in bright and sunny weather they cast ever-changing shadows across the lawn.

Yew is a good grower even on chalky soil. Indeed, the only thing it really detests is waterlogging. Box is less vigorous and in its dwarf varieties can be used to make low hedges for parterres and kitchen gardens. It has fallen out of favour, somewhat, since the prevalence of the disfiguring fungus disease box blight and the box moth, especially with organic growers who are reluctant to use chemical sprays. The spray-on foliar feed 'Topbuxus Health-mix' seems effective at warding off attack. Yew seems to have no such problems. Start with young plants for the sake of economy and they will put on a good 30cm each year if you plant them in decent soil. Train them into cones and pyramids to mark the ends of beds and borders or to mark out a formal vegetable patch.

You can buy them ready-trained and either pot-grown or rootballed (their roots wrapped in hessian) to give instant effect, though the budget will need to be larger. Either way, topiary will lift a garden to new heights. And it might make you smile, too. Clip box in June (around Derby Day) and yew in September, so that it stays crisp in outline through the winter.

# Fruits in store

Late-ripening eating and cooking apples, late-ripening pears, frozen raspberries and other soft fruits from the freezer, plus bottled fruit if you were patient enough to prepare it at harvest time. And jam and marmalade.

# Fruit of the month

## Apple

If you can grow only one tree in your garden, why not plump for an apple? Varieties are available that are grafted onto dwarfing rootstocks such as M.9 (up to 2m) for small gardens, or more vigorous rootstocks like M.25 (up to 6m) if you want that classic apple tree under which you can sit in summer. The rootstock MM106 is somewhere between the two and will make a tree 3 to 4m high. In spring an apple tree is wreathed in pink and white blossom and the fruits follow, cradled in downy-backed leaves. Early varieties ripen in August and later ones are picked in early autumn and stored in a cool, frost-free shed or garage so that they see you through the winter. 'Family Trees' are available with several varieties grafted onto one trunk, but they tend to grow somewhat unevenly.

When it comes to choice of variety, I'd plump for 'Egremont Russet' if you like a rich, nutty favour and don't mind a rough skin, or 'Ashmead's Kernel' if you prefer a rosy tint. 'James Grieve' is wonderfully tangy and has the advantage of being one of the earliest to ripen. 'Cox's Orange Pippin' is the variety everybody knows, but it is prone to disease and not a brilliant choice for gardens. If you want a cooker, 'Bramley's

Seedling' is a whopper but it tends to bear heavily every other year and lighter in its 'off' years. 'Charles Ross' or 'Grenadier' might be a better choice. If you lack the space, grow trained apple trees against a wall or fence in 'espalier' form – with their stems trained out sideways in tiers – or single-tiered espaliers known as 'step-over trees' to edge your veg plot.

# In the kitchen

If you are desperate for crunchy greens (and February is too early to sow salads outdoors) try growing a few sprouting vegetables. You don't need any garden at all. Organically grown seeds of bean sprouts, fenugreek, alfalfa and the like are available in packets. Pour a handful of seeds into a jam-jar and seal the top with a square of muslin or old (washed!) net curtain, held in place with an elastic band. Alternatively, a few pounds will buy you a specially made 'seed sprouter'. Rinse the seeds with water once or twice a day, drain them after each rinsing and stand the jar or sprouter on your windowsill. In just a few days the seeds will have germinated and produced a fluffy mix of seedlings, which are tangy, nutritious and delicious. Use them as a side salad or place them between two slices of freshly buttered wholemeal bread and crunch away!

# Something to read . . .

There are certain books about which I am overly sentimental. *The Tale of Peter Rabbit* by Beatrix Potter is
one of them. It is the first book I remember being read
to me as a child, and I have grown up being able to
recite most of it from memory, first to my children and
now to my grandchildren. The book pays being revis-
ited, not only for its economical prose and to pacify a
fractious child, but also for the astonishingly skillful
illustrations.

Look at the way Beatrix Potter uses ink and watercol-
our to create a living, breathing garden, never mind the
rabbits. Her cabbages are alive, her painting of a variegated pelargonium growing in a terracotta pot a fine
piece of botanical illustration, the white cat watching
the goldfish swimming in the pool a masterpiece of
quiet menace. Granted, Mr McGregor is a mean-spir-
ited son of the soil (it's a wonder I was not put off being
a gardener), but all is well in the end, and the prospect
of bread and milk and blackberries so very comforting.
Funnily enough, I've never really got on with chamo-
mile tea . . .

# Famous gardener of the month

## Christopher Lloyd (1921–2006)

To those who knew him well, Christopher Lloyd was a valued and stimulating – if challenging – friend. His garden at Great Dixter in East Sussex – inherited from his father, the Arts and Crafts architect Nathaniel Lloyd, who wrote *Garden Topiary in Holly, Yew and Box* – became a Mecca for his fans who flocked there from all corners of the earth.

Why? Because he was not only knowledgeable about plants, but had a way of communicating in print his skills, his passions and his prejudices in such a readable and thought-provoking way. He wrote a column for the magazine *Country Life* every week for over forty years, never becoming stale. He believed that a garden should be a living, breathing, vibrant and ever-changing canvas, rather than a patch of land to be preserved in aspic.

'Rules are for breaking,' he said, and at Dixter he was never afraid to experiment with plant combinations and plant breeding, though successful strains would carry the name 'Dixter' rather than 'Christopher Lloyd' – 'I don't want plants named after me,' he said, 'I think that's an awful bit of vanity.' He was funny, grumpy, teasing, curmudgeonly, brave, provocative, and an enemy of what he called 'perceived good taste.' He loved reading; loved music; loved writing letters; loved dachsunds; loved laughing; loved going to Glyndebourne; loved life and had a wide range of friends. When one of them complained that 'All my friends are dying,' he replied, 'Well, get some young ones; I have.'

His books transmit a kind of energy possessed by few other writers. 'Christo', as his friends called him, confessed that he was no garden designer, but his infectious enthusiasm for plants and their placement – influenced by Gertrude Jekyll whom he had met as a child, and his great friend Beth Chatto later in life – rubbed off on those with whom he came into contact, myself included. Read *The Well-Tempered Garden* or *Foliage Plants* and enjoy the well-crafted yet unpretentious prose. His garden at Great Dixter is still open to the public and it remains well worth a visit.

# Seasonality

In a world where instant gratification is 'the norm', something valuable is lost: the enjoyment of anticipation. Today we can buy whatever fruit or vegetable we want, and any flower we fancy for a floral arrangement, at any time of year. Strawberries are available in January, though they will have been flown halfway around the world to get to you and their flavour, you will discover, can seldom match that of those berries warmed by the June sunshine on your own veg patch or allotment.

There is a lot to be said for eating fruit and vegetables in season, and for eating those that are locally produced.

I love parsnips and potatoes, leeks and broccoli in winter and look forward to my asparagus in April, lettuces in May, peas and beans in summer. They offer variety – as well as freshness – when they are picked from the garden or bought from a local farm shop on the land that has actually grown them.

Eating seasonal produce is also a practical way of doing our bit for the environment (how I hate that phrase, though not what it implies) and reducing our carbon footprint – something that was unheard of fifty years ago, but which must now be a part of our thinking process. Eating locally produced food in its natural season is a way of refreshing our palate at the same time as doing our bit for Mother Earth. I know it sounds a grandiose and rather pompous claim, but I do believe in it. And in a couple of month's time, I shall dip my home-grown asparagus in mayonnaise and be glad of the postponement of gratification.

# Wild flower of the month

## Primrose (*Primula vulgaris*)

We wait, we wait and we wait, through December and January and then . . . up it pops, the first primrose of spring. Well, late winter really, but there will be a sheltered, sun-warmed patch where this precursor of another year opens its buds and shows off those flowers of pale buttery yellow that the bees will welcome every bit as much as we do. *Prima rosa* – the first flower of the year – a title that might more accurately be given to the snowdrop, but that is a bulb; the primrose is a 'proper plant'.

It will always be associated with Easter, since its flowers are open in quantity in March and April, but the first ones will usually appear in February, which is why I include it here – as a symbol of hope. It seeds itself freely, and occasional white and pale pink blooms appear – deeper ones, too, where cross-pollination has occurred with the domesticated 'polyanthus'. Primroses proliferate in woodland, in hedgerows and hedgebanks where the soil is moist – they do not much like dry earth, in spite of their willingness to please elsewhere. They remain, due to their bravery, the best-loved harbingers of spring, and the turning of the seasons would be a duller place without them.

*Other local common names*: Early Rose, Pimrose, Butter Rose, Darling of April, Easter Rose, Golden Stars, Simmerin, Buckie-faalie, Lent Rose, May-flooer, May Spink.

# Things you can do

— Knock snow off evergreens before it thaws and breaks branches due to an increase in weight. Take your photos first!

— Provide garden birds with food and fresh water daily.

— Try to finish planting bare-root trees and shrubs this month, though the job can go on until March if you can't get on the earth.

— Plant lily bulbs outdoors and in pots.

— Pot up rooted cuttings in the greenhouse.

— Clean out the greenhouse and wash down the glass – a messy job, but worth it to improve light transmission.

— Prune clematis.

— Try to complete rose pruning.

— Prune autumn-fruiting raspberries, cutting the canes back to ground level.

— If you have a heated propagator you can start sowing tender bedding plants, but make sure you can give them frost-free accommodation until late May.

— Start dahlia tubers into growth in a warm greenhouse if you want to take cuttings from them.

— Divide and replant old clumps of chives.

— Order seeds and seed potatoes from catalogues and online.

— Carry on cultivating the earth on the veg patch, forking out weeds and forking in well-rotted garden compost or manure. Do the job a little and often, rather than in one fell swoop that does your back in.

# Things you should not do

— Don't start sowing seeds under glass unless you are confident you can keep the plants growing in conditions that suit them. It's better to wait a few weeks if you cannot continue to provide heat and light. Later sowings will often catch up.
— Don't plant faded pots of spring-flowering bulbs into the garden at the same depth as they are growing in their pots. They will need to be planted 8–10cm deeper to avoid drying out and to protect them from frost.
— Keep off frosted lawns.
— Don't try to cultivate wet or frozen earth. It will wait.

March

The flowers that bloom in the spring,
Tra la,
Bring promise of merry sunshine.

W.S. GILBERT, *The Mikado*, 1885

How gardeners long for the spring – the start of another
year, and spring must come in March, mustn't it,
because 1st March, or is it 21st, is officially the first day
of spring? Well . . . yes. But nature and the weather is
no respecter of man's idea of a calendar – Gregorian,
Julian or Coptic. Although we may follow the first-
named of these, all manner of vicissitudes can befall our
plans, but nothing can stop us from feeling that we have
turned a corner, and indeed we have in terms of day
length. We will not be fooled, we will not cast a clout,
we will keep open a weather eye, but we will rejoice in
the fact that we are about to slough off the wearying
load of winter and enter a season of movement, of
growth and optimism. (Unless we are a farmer – see
January.)

## Weather

'In like a lion and out like a lamb' goes the saying;
insinuating a degree of predictability that has never
been owned by the month of March. It is a hopeful

aphorism, which assumes that the winds will roar and the tempests will swell between the 1st and the 15th of March, but then things will quieten down to welcome April, the month that most certainly does qualify to be treated as spring. So, at the risk of urging caution and sounding like the voice of doom at the beginning of chapters devoted to the first three months of the year, let me cheer you up by reminding you that the days are longer now, the nights are shorter, and the weather can only get better. I think . . .

# Day length (Skipton)

|  | 1 March | 31 March |
| --- | --- | --- |
| **Dawn** | 06.20 | 06.07 |
| **Sunrise** | 06.54 | 06.41 |
| **Sunset** | 17.46 | 19.43 |
| **Dusk** | 18.20 | 20.17 |
| **Daylight hours** | 10 hours 46 minutes | 12 hours 56 minutes |

**Clocks go forward 1 hour during the early hours of the last Sunday in March.**

# Feast days for gardeners

| | |
|---|---|
| **1 March** | the start of **Meteorological Spring.** |
| **1 March** | **St David's Day** – the patron saint of Wales and the day to wear a daffodil or a leek. |
| **Lent** | begins on the seventh Wednesday before Easter. |
| **17 March** | **St Patrick's Day** – the patron saint of Ireland – a day to wear the shamrock. |
| **20 March** | the **Vernal** or **Spring Equinox** (a time of equal length days and nights). |
| **21 March** | still thought of by many as the **First Day of Spring**. |
| **Easter** | celebrating Christ's Resurrection, 46 days after Lent begins. It is, quite literally, a moveable feast and may fall as early as 22 March. It falls on the first Sunday after the first full moon on or after 21 March. |
| **Last Sunday in March** | **British Summer Time** begins – the clocks go forward one hour. (Remember |

the mnemonic: 'Spring forward, fall back.) You lose an hour in bed but, what the heck, the mornings are lighter and brighter and the birds are singing!

**Mothering Sunday** usually falls in March. It is celebrated on the fourth Sunday of Lent, three weeks before Easter. It is traditionally a day when mums are presented with a posy of flowers by their children. The American name of 'Mother's Day' is frequently used instead. Hmmm . . .

# Tree of the month

## *Winter-flowering cherry (Prunus* x *subhirtella* 'Autumnalis Rosea')*

I know it's March, and I know part of its name is 'Autumnalis', but this dainty flowering cherry keeps on producing its double pale pink blossoms on bare branches from winter right into spring and is often a glorious sight in March.

One of its greatest attributes is that its structure is not overpowering. The canopy is relatively thin (the leaves are quite small for a cherry) and the tree does not cast much in the way of shade. Train a clematis such as a variety of *Clematis viticella* or *C. texensis* up through its lower branches for summer colour and chop down the

clematis to ground level in winter, removing the dead
stems and foliage from the cherry's canopy. The clematis
won't mind at all, and neither will the cherry, and that
way you get two flower displays from the one space.

Height and spread after 20 years: 4–5m.

# Music to listen to . . .

Trying to listen to music and write at the same time I
find extraordinarily difficult, since splitting my attention
between the two results in a diminution of pleasure on
both counts. As a result I write in silence and listen to
music when I am pottering about with books or plants – I
seem to have a lot of both. When it comes to the music, I
tend to have two or three pieces that I play over and over
during the space of a week and then, as I tire of their
familiarity, I play something completely different.

One album I regularly come back to is the soundtrack
to the film of *Pride and Prejudice* composed by Dario
Marianelli. The opening movement – Dawn – is so
fitting for spring since it conjures up a kind of re-
awakening. It also conjures up glorious images of
Groombridge Place in Kent, which took on the role of
Longbourn, and of Elizabeth Bennett as played by
Kiera Knightley (rather neat that her surname is shared
with Jane Austen's hero in *Emma*). Enjoy *P and P* as a
change from Vivaldi's *Four Seasons – Spring* – and
Christian Sinding's *Rustle of Spring*, which would both
have been far too obvious to recommend.

# Vegetables from the garden this month

Sprouting broccoli, cabbage (sown last summer), late cauliflowers, kale, leeks, rhubarb, spinach.

# Vegetables in store

Jerusalem artichokes, beetroot, carrots, onions, parsnips, potatoes, shallots, swedes, turnips.

# Something to muse upon . . .

For me, the images conjured up by a poem add immeasurably to the music of the words, the spirit evoked and the sentiment expressed. *The Lake Isle of Innisfree* adds a third dimension – the *sounds* of nature. In those moments when we feel a need to escape to a place of sanctuary, where nature can work her magic, this is the poem that hits the spot:

I will arise and go now, and go to Innisfree,
And a small cabin build there, of clay and wattles made:
Nine bean-rows will I have there, a hive for the
   honey-bee;
And live alone in the bee-loud glade.

And I shall have some peace there, for peace comes
    dropping slow,
Dropping from the veils of the morning to where the
    cricket sings;
There midnight's all a glimmer, and noon a purple glow,
And evening full of the linnet's wings.

I will arise and go now, for always night and day
I hear lake water lapping with low sounds by the shore;
While I stand on the roadway, or on the pavements grey,
I hear it in the deep heart's core.

W.B. YEATS (1865–1939), *The Lake Isle of Innisfree*

# Flower of the month

## Helleborus hybrids (Lenten hellebore)

That forest of leaves – great shiny, evergreen umbrellas
that erupt in spring and tower over the clumps through
the summer – hide what is happening below in late
winter. Scissor them off at ground level in January and
you will see great crooks of stems unfurling from the
ground to reveal their glorious saucer-shaped flower
clusters of white, yellow, green, pink, crimson, slate-
purple and a host of shades in-between, the paler hues
spotted with crimson. There are double-flowered
strains, too, and they are all brilliant for growing in
semi-shade among trees or shrubs where early spring
flowers will brighten your life.

The clumps will expand over the years, and the plants will seed themselves freely, so if you want to grow and select your own strains, you will not find it at all difficult. Most are hybrids of *Helleborus orientalis* and there are strains like the 'Ashwood Garden Hybrids' raised by that excellent plantsman John Massey in the Midlands. Lenten hellebores begin flowering in January and will continue through to April in any reasonable soil. As plants go, they are about the most obliging of the lot.

Problems? Removing the leaves in January each year will help to reduce the incidence of a disease known as 'black death'. It happens rarely and infected plants that show signs of black areas on the leaves and flowers are best disposed of completely. Mice can nibble at the emerging flower buds if you allow the leaves to remain in place for too long offering them cover. Hellebores tend to wilt if you pick them. Instead, float the individual flowers in groups on a wide bowl of water in the middle of a table so that they face upwards and you can admire their shape and colour.

Height and spread: 30cm.

# Flowers in the garden

Anemones, Bergenia (elephant's ears), Camellias, Chaenomeles (Japanese quince), *Cornus mas*, Corylopsis, Crocuses, Doronicum (leopard's bane), Daphne, Erica (winter-flowering heathers),

Erythronium (dog's tooth violet), Forsythia, Gentiana (gentians), Helleborus (hellebores), Hyacinths, Ipheion, *Iris unguicularis* (winter-flowering iris), Leucojum (snowflake), *Magnolia campbellii*, *Magnolia stellata* (star magnolia), Mahonia, Narcissi, *Parrotia persica* (Persian ironwood), Pieris, Primulas (including primroses and polyanthus), Prunus (flowering almond, winter cherry), *Prunus persica* (Peach), *Prunus spinosa* (blackthorn), Pulmonaria (lungwort), Ranunculus (celandine), early Rhododendrons, Ribes (*flowering currant*), Salix (willow), Saxifraga (saxifrage), Scilla, early Tulips (*Tulipa kaufmanniana*), *Viburnum bodnantense, V. burkwoodii, Viburnum tinus*, Violas.

# Herbs from the garden

Bay, chives, first sprigs of mint, rosemary, sage, thyme plus basil and coriander indoors.

# A garden to visit in March

Spring comes earlier to the south-west of England than it does elsewhere on the mainland, and there are any number of gardens in Devon and Cornwall that are worth visiting for camellias, rhododendrons and magnolias in particular. I am especially fond of Antony, at Torpoint, which has wonderful views over the Tamar and woodland packed with all three of these spring beauties.

You could make a weekend of it, sampling the local scenery, visiting a clutch of gardens and emptying the back of the car so that you have time to shop at the likes of Burncoose Nurseries, having explored the amazing gardens at Caerhays Castle, near St Austell in Cornwall. *Camellia* x *williamsii* was bred here in 1923 and the Williams family still preside over the estate and the nursery, which offers an amazing range of hardy plants and a good number of slightly tender exotics. The finest gardens of the West of England are nothing short of breathtaking between March and June.

# Allotments

It was the Allotments Act of 1887 that made provision for land to be acquired by local authorities for the purpose of allowing householders to grow their own food, which led to the creation of what we call 'allotments'. The Smallholding and Allotments Act of 1907 forced local councils to provide land for that purpose if there was a demand from more than six people. London – where spare land is in short supply – was made an exception.

At the time of the First World War in 1914, there were around half a million allotments in England and by 1917, with the need to grow more food at home, that number had risen to over one-and-a-half million.

Numbers declined after the 1914–18 war, when much of the land that had been requisitioned to grow food was handed back. By 1930 only a million remained.

At the beginning of the Second World War in 1939, there were just over 800,000 allotments, but with the encouragement to 'Dig For Victory' another half million plots were added. Even London's parks were dug up to grow food. The end of the war saw a massive decline in numbers until, by the 1970s, only half a million plots remained. By the turn of the century that number had been reduced to just over a quarter of a million, in spite of a sharp resurgence in the mid-1970s when *The Good Life* was transmitted on television and the 'Grow Your Own' movement had a brief resurgence.

Today, with renewed interest in home-grown fruit and veg, and the importance of knowing where our food has come from, numbers of council allotments hover around the 300,000 mark with 100,000 people on a waiting list, according to the National Society for Allotment and Leisure Gardeners. Council allotments continue to provide a valuable resource for gardeners who lack the space to grow food and veg at home. They have become an important catalyst in the coming together of many different cultures to grow all kinds of food from the basic and mundane to the exotic. Flowers, too, brighten up these patches and the homes for which they are grown.

Where would we be without allotments? Camaraderie, gentle rivalry and the passing on of growing skills are all fostered by this simple but highly valued use of land across the country.

# Bird of the month

## Blackbird (*Turdus merula*)

Of all the songs in all the land, the blackbird's evening recital is the most treasured in our house. There he sits on the chimneypot (Mrs Blackbird is not such a show-off) and accompanies both the rising and the setting of the sun between late January and July. The strident 'chook-chook-chook' alarm call is heard whenever the bird is disturbed by human or cat. Black of coat and orange of bill, the blackbird is easily recognisable; as is the female with her dark brown plumage and less colourful beak. The oldest blackbird recorded reached an impressive twenty years and the species is our most numerous native bird.

The female alone incubates the eggs while the male takes on feeding duties. There are usually four or five pale blue eggs laid in two or three broods between March and May, with eggs hatching between twelve and twenty days and the youngsters fledging in just a couple of weeks. Insects, worms and berries are their food (berries being especially important through the winter) and although blackbirds will land on a bird table, they prefer to feed on the ground. There are around 5 million pairs in Britain.

# Feeding beds and borders

There is a moment in early spring when I think, This is it! – the moment when my beds and borders, lashed by rain for months on end, are ready to have their food reserves replenished in the form of fertiliser. There is no point in applying plant food too early – it will be leached through the soil before the plants have woken up and are ready to use it. By March, root growth will have begun and that's when I give all my border plants, shrubs, roses and the like, a good sprinkling of blood, bone and fishmeal.

Why that particular fertiliser? Because it contains all three main plant nutrients – Nitrogen (N), Phosphates (P) and Potash (K). What's more, being organic, it needs soil bacteria to break it down and make it available to plants. (Inorganic fertilisers such as Growmore are capable of being absorbed as soon as they are in solution, which means that soil bacteria become redundant.)

Two or three clenched fistfuls to the square metre is about right and then lightly fork it in. I give a second dose of the feed in June to keep plants growing well. Spreading a 5-cm thick mulch of bark or well-rotted garden compost after the spring fertiliser application will help seal in moisture and keep down weeds. Blood, bone and fish is a good lawn feed, too. Applied in late March, it will release its nutrients more slowly than inorganic lawn feeds and is less likely to lead to scorching of the grass, though with all fertiliser applications, the best timing is just before rain is forecast so that the food can be washed into the earth as soon as possible.

# Fruits in store

Late-ripening apples and pears, bottled fruit and frozen soft fruits. And there's always Sloe or Damson Gin to warm you up on a cold spring day! Oh, and jams and marmalade, of course.

# Fruit of the month

## Jam

A bit of a cheat, I suppose, but we are reaching the end of the season for stored fruits – apart from those that are bottled or frozen. It's all too easy to gorge ourselves on fresh fruit with little thought given to the end of the season when they run out; we've had a long wait for them, after all. A pot or ten of jam made during the summer is like bottled sunshine in winter and early spring. Hugh, for whom this almanac is dedicated, kindly keeps me supplied with marmalade made in December, when the Seville oranges appear in the shops. Home-grown strawberries (if you have enough) and raspberries (you *will* have enough) along with damsons and plums make great jam in summer and there is no shortage of recipes in books, from grannies and online.

Strawberry jam spread on a freshly baked scone with a dollop of clotted cream, accompanied by a warming brew of Yorkshire Tea or Earl Grey, and you have the perfect spring afternoon pick-me-up. Mouthwatering!

# Something to read . . .

As the celandines and primroses open, we are reminded of the floral riches of our islands as the season progresses, from these early risers to the late summer heathers of moorland and the marjoram that grows in wild flower meadows. There are many British Floras (the name given to wild flower dictionaries) and I treasure a fair number of them, from the Rev. W. Keble Martin's *Concise British Flora in Colour*, with its delightful watercolours, first published in 1965 and given to me five years later by my auntie for my twenty-first birthday, to Sarah Raven's *Wild Flowers*, with photographs by Jonathan Buckley, published in 2011, which is elegantly written and mouthwateringly illustrated. The wild flowers are divided into habitats to make locating them fairly easy, and there is a comprehensive index.

But if I had to choose just one British Flora that combined beauty with usefulness, it would be Rae Spencer-Jones and Sarah Cuttle's *Wild Flowers of Britain and Ireland*, whose photographs, arranged according to colour, make identification easy.

# Famous gardener of the month

## Tom Stuart-Smith (1960–)

There are hundreds of 'garden designers' in Britain. The best ones not only know about design but also have an in-depth knowledge of the plants that go into making them. Tom Stuart-Smith is one such, as his designs across the country testify. He's a personable soul, quite intellectual, very tall, seldom wears a tie and is blessed with a fine sense of humour and the ability to communicate as well as having a talent for putting plants together.

Take a look at the new RHS Garden Bridgewater at Worsley near Salford, Greater Manchester, or the renovated and replanted rock garden at Chatsworth in Derbyshire – both show Tom's mastery of the plantsman's art. He designed a new garden at Windsor Castle for The Queen's Golden Jubilee and the gardens around the Bicentenary Glasshouse at RHS Wisley.

Having trained with Hal Moggridge and Elizabeth Banks, Tom established his own garden design business in 1998, since when he has never been short of commissions. He is not simply a repeater of old ideas and traditional plans, but an originator whose contemporary designs make sense. You can see his work occasionally at the Chelsea Flower Show and read his two-volume work *Drawn From the Land* (Thames and Hudson, 2021), which is replete with plans and explanations. His reputation as one of the country's finest landscape architects and garden designers is well deserved.

# Sowing seeds under cover

At last the moment is here. It's OK for those gardeners with heated greenhouses who can start sowing seeds in January and February, knowing that they can offer the resulting young plants protection until the time comes to put them out in the garden, but heat costs money and sometimes funds don't stretch that far. In March, if your cool or even unheated greenhouse can be fitted with a small, heated propagating case, you can start sowing seeds, knowing that by the time the seedlings are pricked out into seed trays of compost in April and the young plants are growing away well, the unheated structure will probably offer sufficient protection to ensure their survival.

It's always tempting to sow too early. Tomatoes in particular will literally turn blue with cold if you start them off too soon and cannot provide at least 12–15°C. Better to wait until outdoor temperatures are that bit higher and the days are longer, for light is every bit as important as heat to seedlings – if they have too little of it they will become tall, drawn and spindly, and their tender tissues will fall prey to fungus diseases more readily than when they are sturdy, stocky and bristling with health.

If you can invest in a plastic propagating case with a heated base, the seeds will have the 'bottom heat' they need to encourage germination. If your case can be equipped with overhead lighting (simple kits are available that will not break the bank), then the problem of light intensity is also addressed. The larger the propagating case, the more pots and trays you will be able to fit into it, and a thermostatically controlled case will make sure you don't break the bank. I wouldn't be without mine – it still excites me to check the pots and trays each morning to see what has emerged overnight. Simple pleasures . . .

# Wild flower of the month

## Cowslip (*Primula veris*)

Once they were everywhere; now we rejoice when we see them, for they are not nearly so frequently encountered as the primrose thanks to the ploughing up of ancient pastures and the over-use of chemical herbicides in farming. But they are coming back; you'll see country roundabouts as yellow as a field of oil-seed rape (though gentler of hue) and on chalk downland they can be found in abundance. My own Hampshire wild flower meadow, sown by hand with a chalk downland seed mixture around fifteen years ago, now erupts with thousands of them each March and April. They seed themselves (the meadow is not cut until early September, by which time the seeds have fallen) and increase in number year on year.

From a rosette of crinkly primrose leaves arises a 20-cm high stalk carrying anything from ten to thirty individual five-petalled flowers. The rich yellow blooms, each one delicately blotched inside with tiny orange spots, nod gracefully and the pale green calyx in which each flower sits is long and goblet shaped. The cowslip is a flower of the meadow and the brightly lit hedge-bank, where the soil is alkaline or perhaps slightly acid. It loves the sunshine and fizzles out in woodland. The oxlip (*Primula elatior*) is similar, but the petals are wider and more primrose-like. Kneel down on a still, sunny day to enjoy the gently fruity fragrance and observe how the bees love the flowers.

*Other local common names*: Bunch of Keys, Cove-keys, Cowslap, Peggles, Paigle, Culverkeys, Fairy Bells, Fairy Cups, St Peter's Keys, Cowslop, Lady's Fingers, Herb Peter, Long Legs, Freckled Face, Golden Drops, Cow's Mouth, Cow Strupple, Racconals.

# Things you can do

— Divide border perennials (especially on heavy soils where autumn division might have led to rotting).
— Complete rose pruning.
— Mulch beds and borders with chipped bark or well-rotted garden compost after sprinkling an organic fertiliser on the surface and forking it in.
— Lay turf to make new lawns.
— Take cuttings of greenhouse plants.
— Pot up rooted cuttings.
— Sow bedding plants in a propagator.
— Prick out seedlings as soon as they are large enough to handle.
— Sow sweet peas outdoors in a mild spell.
— Repot houseplants.
— Sow those vegetable and herb seeds that need to be started in a greenhouse.
— Plant strawberries.
— Sow lettuce seeds in a cold frame or greenhouse.
— Cut hard back dogwoods grown for their coloured stems.
— Outdoor veg sowings can be made if the weather forecast is promising.
— Sow hardy annuals outdoors . . . if the weather forecast is promising.
— Plant gladioli.
— Plant asparagus crowns in prepared soil.
— Start dahlias into growth in a greenhouse.

— Take dahlia cuttings and root in a propagator.
— Start begonia tubers into growth.
— Plant early potatoes if the soil is not too wet or frozen.
— Plant snowdrops 'in the green' and dig up, divide and replant overcrowded clumps if you want to increase your plantation.
— Clip over ivy on walls to reduce its weight.

# Things you should not do

— Don't risk sowing things outdoors unless you have very well-drained soil and are reasonably confident that a prolonged spell of severe weather is not on the way. (I know, it's tricky, so if in doubt, hold back . . .)
— Don't stop feeding garden birds and providing water.
— Don't walk on frosted grass (you'll see your footprints when the thaw comes).
— Don't lose heart; the spring proper will soon be here.

April

April the angel of the months, the young
Love of the year.

VITA SACKVILLE-WEST, *The Garden*, 1946

It is here at last. The proper spring month, which is where I go into my cautionary tale of the weather being treacherous – warm and sunny one minute, frosty the next – the bane of all spring blossom and a dispiriting kick in the pants for gardeners who thought, too soon, that winter was over and done.

But that's it. I will be negative no more. Instead, I will think of those glorious April days and believe with Sissinghurst's chatelaine Vita Sackville-West that April is 'the angel of the months', even if she can be a bit of a devil from time to time. No! I said I would not . . .

Shoots are pushing up from the ground, buds are bursting open on fruit trees, all is optimism and Nature comes at us with a rush in field and hedgerow, garden and arboretum. There is now so much to see, so much to keep up with, for April is the month in which the rate of growth in the garden goes up a gear. In February and March we could move faster than Nature, but now she has slipped into overdrive and we must try to match her pace. That does not mean going into a flat panic, but pottering out there a little and often and taking time to enjoy the blossom and the bulbs erupting from damp yet sun-warmed earth.

The air smells different now; the scent of growth and renewal is a heady brew that lifts the spirits after those dark, dank months of slumber. The garden is awake; thank God for spring!

# Weather

'April showers bring May flowers' is the old saw, and, indeed April can be a wet month. But it can also be surprisingly dry, and a dry April will often affect the flowering of spring bulbs the following year, since daffodils and narcissi are producing next year's flowers down in the bulbs when this year's blooms are opening. They like a reasonable amount of moisture. The real treachery comes from late frosts that strike when apple, pear and plum blossom is in full fig, for then the chilly air can blacken the centre of each flower and prevent fruit setting. With any luck we will escape such disasters. We live in hope at any rate . . .

# Day length (Skipton)

|  | 1 April | 30 April |
|---|---|---|
| **Dawn** | 06.04 | 04.52 |
| **Sunrise** | 06.38 | 05.31 |
| **Sunset** | 19.45 | 20.39 |

| Dusk | 20.19 | 21.18 |
| --- | --- | --- |
| Daylight hours | 13 hours exactly | 15 hours 1 minute |

# Feast days for gardeners

**1 April**  **All Fools' Day** – but play your tricks before noon or the joke is on you.

**Palm Sunday**  the Sunday before Easter – often falls this month. Celebrating Christ's entry into Jerusalem and the strewing of palm leaves in his path by way of praise and acknowledgement that he was the King of Israel. Crosses made from palm fronds are distributed in churches of the Christian faith.

**Good Friday**  the Friday before Easter – the day Christ was crucified. It is said that this is the only day in the gardener's calendar when parsley seeds do not go 'nine times to the devil' before they germinate. Well, my mum swore by it . . .

**Easter Day**  celebrating the risen Christ. A time for hiding eggs in the garden for children and grandchildren to discover.

83

| | |
|---|---|
| **14 April** | **St Tibertius' Day** – traditionally the day on which the cuckoo begins to sing (until St John's Day on 24 June). |
| **14 April** | **National Gardening Day** – a time to celebrate things that grow and encourage others to do the same. |
| **23 April** | **St George's Day** – the patron saint of England. |
| **23 April** | start of the British asparagus season (ends on 21 June at the Summer Solstice). |

# Tree of the month

## Magnolia

You can see them in March, daring to split open the downy sheaths that protect the cup-shaped waxy goblets of flowers through the winter. In cities, where the air is warmer, the earliest magnolias can be seen flowering in that month, but such earliness worries me, for the glorious blooms are susceptible to frost damage that, in a bad year, will leaves the trees covered in caramel-coloured rags rather than the white, pink or purple flowers that nature – and man – intended. But they are spared often enough to make planting them worthwhile.

The flowers are narrow-petalled and starry on the earliest to bloom: the star magnolia, *Magnolia stellata*, which has a delightful pink form 'Leonard Messel'. The varieties of *M*. x *soulangeana* are by far the most common and the most spectacular with large goblet-shaped flowers carried on bare branches – the leaves emerge later.

Look for the varieties 'Rustica Rubra' and 'Lennei', which are generously flushed with pink on the outside of the flower, or 'Alba Superba', which is white. They prefer lime-free soil and a sheltered spot, and should be planted where they can be grown without being pruned. Chop at them and they look butchered. Try not to disturb the fleshy roots at planting time, since to do so can delay flowering. For magnolia blooms later in the year seek out *M. wilsonii*, which has much a wider-faced white flower with a central boss of burgundy stamens. Mouthwatering!

Height and spread: *M*. x *soulangeana* 6m; *M. stellata* 1.5–2.5 x 5m.

# Music to listen to . . .

There are certain pieces of music that lend themselves perfectly to this time of year. *On Hearing the First Cuckoo in Spring* by Frederick Delius is one such, and it will fill you with renewed hope, as will *The Lark Ascending* by Ralph Vaughan Williams, now so popular that it regularly achieves the No. 1 spot in the Classic

FM 'Hall of Fame'. Listen to it played by the violinist Iona Brown, who chose it as the very last piece she played at the end of her all too short career. For lighter fare, enjoy *The Water Mill* by Ronald Binge, and picture yourself in some bucolic idyll where the mill-wheel gently plays. Which leads me neatly on to Schubert's *Piano Quintet in A* – The Trout. And if these don't put you in the mood for spring, nothing will.

## Vegetables from the garden this month

Asparagus, sprouting broccoli, summer-sown cabbage, kale, leeks, lettuce, spring onions, rhubarb, spinach.

## Vegetables in store

Jerusalem artichokes, beetroot, carrots, onions, parsnips, potatoes, shallots, swedes, turnips.

## Something to muse upon . . .

There is one poem that I regularly recite from memory at this time of year, or a part of it at least. A.E. Houseman was born in Bromsgrove in 1859. He was a classical scholar as well as a poet and *A Shropshire Lad*, published in 1896, remains his best-known work. I can wander among my flowering cherry trees in April – the most beautiful variety of all 'Shirotae' with fully double white blossoms – and say out loud:

Loveliest of trees, the cherry now
Is hung with bloom along the bough,
And stands about the woodland ride
Wearing white for Eastertide.

Now, of my threescore years and ten,
Twenty will not come again,
And take from seventy springs a score,
It only leaves me fifty more.

And since to look at things in bloom
Fifty springs are little room,
About the woodlands I will go
To see the cherry hung with snow.

A.E. HOUSEMAN (1859–1936), *A Shropshire Lad*

It brings a tear to my eye just to set down these words. The emotion while saying them out loud as I duck under cherry boughs weighed down with blossom is almost overpowering. I know, too romantic for my own good . . .

# Flower of the month

## Tulip

Few flowers are as exciting to watch develop as the tulip. Planted late in the autumn – around November – the smooth, pointed bulbs sit in the earth, plumping up and readying themselves to break the surface of the soil in February or March. The wide, rubbery leaves cradle the flower bud between them, but it is invisible at first, gradually emerging as the stalk extends but remaining plain green. Slowly the outer tepals show their true colours and that glaucous green is gradually stained with red, or yellow or pink or white as the stalk continues to push upwards. The shortest tulips – the true species, which are native mainly to those countries ending in 'stan' – are a few inches high and grown in pots or rock gardens.

It is the hybrids that are the stalwarts of the English garden, with the earliest flowering varieties of *Tulipa greigii* and *T. kaufmanniana* growing to under a foot, and the tallest Darwin hybrids to double that in late April and early May. Plant them in pots and in beds and borders. If you want them to come up in the garden year after year, without having to dig them up and dry them, which was usually the custom, plant them a good 20cm deep. Some varieties, such as 'Spring Green', will happily perpetuate given this treatment, others will fizzle out – as will they all when planted in lawn or meadow where they do not naturalise as readily as daffodils and narcissi, thanks to competition from the grass.

There are tulips in every colour except blue. The Parrot types, beloved of the Dutch masters, have petals that are frilly edged, the Rembrandt types are streaked with contrasting colours, and the Lily-flowered tulips have blooms that are nipped in at the waist and whose pointed petals curl gently outwards with supreme elegance. Treat yourself to a few new varieties every year and experience the thrill of 'Tulipomania', which was rife in 1634, when single bulbs changed hands for unconscionable amounts of money. Now they can be had for less than 50p apiece. Bargain!

# Flowers in the garden

Acer (maple), *Akebia quinata* (chocolate vine), Alliums, *Alyssum saxatile*, Amelanchier (snowy mespilus), Anemones, Arabis, Arbutus (strawberry tree), Aubrieta, Azalea, Berberis, Bergenia (elephant's ears), Caltha (kingcup), Camellia, Ceanothus (Californian lilac), Chaenomeles (japonica or cydonia), Cheiranthus (wall-flower), Chionodoxa, Choisya (Mexican orange blossom), *Clematis alpina*, Convallaria (lily-of-the-valley), Cytisus (broom), Daphne, Dodecatheon (shooting stars), Doronicum (leopard's bane), Epimedium, Erica (heather), Erythronium (dog's tooth violet), Euphorbia, Forsythia, Fritillaria, Gentiana (gentians), Geum, Helleborus (hellebores), Heuchera, Hyacinthus (hyacinths), Iberis, Ipheion, *Iris japonica*, Magnolia, Mahonia, Muscari (grape hyacinth), Myosotis (forget-me-not), Narcissus, Omphalodes, Osmanthus, Pieris, Polygonatum (Solomon's seal), Primula, Prunus (flowering cherry), Pulmonaria (lungwort), Pulsatilla, Pyrus (pear), Ranunculus, Rhododendron, Ribes (flowering currant), Saxifraga (saxifrage), Scilla (squill, bluebell), Skimmia, *Spiraea arguta* (bridal wreath), Trillium, Tulipa (tulip), Viburnum, Vinca (periwinkle), Viola.

# Herbs from the garden

Bay, chives, mint, rosemary, sage, thyme, plus basil and coriander indoors.

# A garden to visit in April

So many gardens are coming to life this month that it should not be difficult to find somewhere local to see rhododendrons and camellias coming into their own, beds and borders stuffed with tulips, and woodland floors carpeted with bluebells. This is such an exciting time of year as the earth really does erupt with flowers. I am an enormous fan of Chatsworth in Derbyshire, where the garden boasts not only woodland and more formal areas, but such spectacles as the Cascade – designed by Louis XIV's engineer Monsieur Grillet and completed in 1696 – where the water flows from a domed folly down a long series of steps.

The newly reconstructed rock garden by Tom Stuart-Smith has transformed a monolithic rocky valley into a floral wonder, and the Emperor Fountain, installed to impress Tsar Nicholas I of Russia (who never came to see it) shoots 6om into the air. The kitchen garden and the greenhouses where the most delicious grape in the world – 'Muscat of Alexandria' – is grown will make any 'grow your own' enthusiast green with envy. The garden and estate are large enough – and spectacular enough – to maintain the interest of visitors young and old for an entire day. Keep an eye open for special events throughout the year – the Country Fair in early September is especially enjoyable, and at Christmas the house and gardens are magically enhanced with sumptuous decorations and illuminations.

# Rock plants and alpines

Rockeries – or more properly, rock gardens – have rather gone out of fashion, which is certainly a shame as far as miniature alpine plants are concerned, not least because many of them flower early and brighten the garden before summer borders have really got going. That said, it is understandable that 'dog's graves' and 'plum puddings' (as the great Yorkshire alpine gardener Reginald Farrer called those rock-studded mounds of grey earth that look as though they are nothing more than spoil heaps, especially when sited alongside a garden pond) have fallen from favour.

Well-constructed rock gardens, with handsome chunks of stone whose strata are carefully arranged to look natural, can be attractive garden features, but they don't really fit into small modern gardens. And yet that's no reason to stop growing alpines. They are blissfully happy in stone sinks (pricey) or even porcelain ones that can be treated with a bonding agent and then coated with a moistened mixture of cement, soil and peat-free compost, which will harden to look like stone.

Alternatively, the really precious alpine treasures can be grown in terracotta pots and shallower 'pans' in an unheated greenhouse, where they can be protected from the weather in their season of flowering and stood outdoors for the summer. All they demand is good light, gritty, well-drained compost and a surface mulch of fine grit to prevent them being mud-splashed, along with protection from winter wet.

That said, the hardier alpines like aubrieta, arabis and *Alyssum saxatile* are dead easy to grow and look spectacular when planted in low retaining walls, where their hummocks of foliage tumble down the sides of the structure and are smothered in flowers each spring. So think twice before deciding that alpines and rock plants are not for you. Growing a variety of them in pots and pans can be as addictive as stamp collecting and a darned sight more spectacular – and satisfying, if you can fulfill their modest needs.

# Bird of the month

## Barn swallow (*Hirundo rustica*)

Oh, how we long for them to return! Sometimes they do; sometimes they choose another place to build their cup-shaped nests of grey mud rather than the lofty corrugated iron lean-to that we fondly call our barn. Red-faced, blue-black-backed with a white undercarriage, the sound of their chit-chit-twittering as they arrive in April makes them sound as excited as we are. Their long, forked tail streamers allow us to distinguish them from the shorter-tailed house martin in flight and their colouring is altogether more sleek in its navy blue, compared with the martin's dusky brown.

How is it possible that this slender bird has travelled to us from South Africa? Up across the Sahara Desert they fly, through Morocco then Spain, across the Pyrenees and western France before beginning their descent into Britain. Come September the route will be reversed, as they leave our autumnal temperatures to warm their bones and feathers in Africa. Most colonies return to the same place as in previous years, and the nests are frequently re-used, often for ten or fifteen years; the record being forty-eight years, which says a lot for the swallow's skills as an architect.

The birds themselves pair for life, even if the males are somewhat free with their affections, sometimes having two mates. Most swallows live for around three years. They will have one, two or even three broods on

93

occasion, depending on the season, each clutch of eggs numbering between three and seven. The eggs will hatch in a couple of weeks and the youngsters will fly after about three weeks, often landing on the ground, which makes them vulnerable where domestic cats are concerned. Swallows feed on insects throughout the spring and summer, and when the time comes for them to depart our shores in September, they will often congregate on telephone wires, chuntering to each other about . . . what? The weather? The likelihood of a delayed departure? One day they are there and the next day they are gone. Until next year . . .

The UK migratory population is thought to be around 500,000 pairs.

# Weeds

One person's weed is another person's wild flower and, as any wise old gardener will tell you, the true definition of a weed is 'any plant growing where it is not wanted'. Those who dislike gardening imagine that the pursuit is all about weeding, and yet removing weeds from my own garden occupies less time than almost anything else. Why? Because I plant things close together in soil that has been beefed up with well-rotted garden compost, and that means there is little room for weeds to grow between the cultivated plants. Any that do succeed are easily prised out. When new beds and borders are to be made in really weedy ground, it should be forked over and any thick-rooted perennial weeds removed.

Easy to say, I know, and a back-breaking job if tackled hastily. Take your time. You could treat the patch of ground with weedkiller a few weeks before cultivating it, but if you want to garden organically there are more environmentally friendly ways of achieving the same ends. If you run a mower over the ground (having cleared it of stones and the like beforehand) and you do this once a week for an entire season, you'll find that many weeds – like ground elder, bindweed and horsetail – are killed off anyway. They cannot stand repeated cutting back.

Alternatively you can cover the ground with thick black polythene weighted down with bricks and leave it in place for a year. The trouble is it doesn't look very nice. Better perhaps to tackle the ground with a fork a little at a time and be assiduous about pulling out the thick weed roots. If you are cutting out a new bed in an existing lawn the problem is not nearly so daunting, for dandelions are about the only thick-rooted weeds that can withstand mowing.

Once you've forked over the ground (or dug it if you are happy to wield a spade), enrich the earth with well-rotted compost or manure and then plant it up, remembering not to leave massive gaps. Yes, the cultivated plants need enough room to grow, but . . . well . . . use your discretion. Water them in and then mulch the ground with a 5-cm deep layer of chipped bark to seal in the moisture and prevent weed seeds from landing and germinating on the surface. Within a year – two at most – your beds and borders will be so full of the plants you love that there will simply be no room for weeds. They'll wisely decide to make a home in some-one else's garden. One with large gaps between the cultivated plants . . .

## Fruits in store

Late-ripening apples and pears, anything you've put in the freezer, jams and marmalade.

## Fruit of the month

### Rhubarb

Yes, I know it's really a vegetable, because it doesn't contain seeds or pips, but we always think of rhubarb, whose euphonious Latin name is *Rheum rhaponticum*, as a fruit because of its flavour. The wild plant is native to the mountains of south-western Bulgaria, but most folk think it comes from Yorkshire. Why? That's where you'll find 'The Rhubarb Triangle': around 23 sq km between the towns of Wakefield, Rothwell and Morley where the majority of early forced rhubarb is grown. It is picked by candlelight to avoid flooding the darkened sheds with light and so compromising the tenderness of the fluorescent pink stalks that are drawn up by the darkness.

In the garden, rhubarb likes a really rich soil and will come into season in the open from March onwards in most years, but it is at its most vigorous and tender in April, carrying on cropping until June, after which it can get a bit tough. Allow the plant to regain its strength after that month, pulling off any leaves and stalks that start to rot. You can bring it on as early as March if you cover it with a terracotta rhubarb forcer in January, though my granddad used an upturned bucket filled with straw on his allotment. The darkness and shelter encourages early growth and the sticks are delicious and tender. Remove the forcing pot (or bucket) when you have picked enough for a few pies or crumbles, to allow the plant fresh air and light.

There are umpteen varieties – some red-stalked and some green – from the old favorites 'Prince Albert', 'Holstein Blood Red' and 'Timperley Early', which will sometimes crop in February when forced, to the flavoursome 'Hawke's Champagne'. Look out, too, for newer varieties such as 'Chipman's Canadian Red', 'Polish Raspberry' and 'German Wine'.

# Something to read . . .

If you have always dreamed of having a secret garden, then you have probably read the book or seen the film or the television series based upon it. Frances Hodgson Burnett published *The Secret Garden* in 1911 and since then it has been a perennial favourite of gardeners and non-gardeners alike. It tells the story of Mary Lennox, orphaned in India and sent to live with her uncle at Misselthwaite Manor in Yorkshire. When she arrived, 'everybody said she was the most disagreeable-looking child ever seen.' But Mary, who stumbles upon the key to a long-neglected walled garden and is led by a robin to its ivy-covered door, is transformed by the overgrown wilderness – along with Colin, the sickly child of the household and her new-found friend Dickon – into a child of nature:

> It was the sweetest, most mysterious-looking place anyone could imagine. The high walls which shut it in were covered with the leafless stems of climbing roses, which were so thick they were matted together. Mary Lennox knew they were roses because she had seen a great many roses in India. All the ground was covered with grass of a wintry brown, and out of it grew clumps of bushes which were surely rose bushes if they were alive . . .

I leave it to you to read on.

# Lawns

Do not be put off having a lawn by all those who think that re-wilding is the answer. There is room for all types of grass in a garden – whether as ornamental plants in beds and borders, as a wild-flower filled sward (perhaps around the edge of the lawn) or as a close-mown patch that pleases you with its neatly parallel stripes.

Moderation in all things is the answer, and if you care about wildlife, remember that blackbirds, thrushes, starlings and woodpeckers will love to peck at your lawn for worms and insects, while other birds remove the moss for their nests. They don't care too much for long grass. Lawn care? There is no need to go mad. Rake out dead grass (thatch) and moss in March or April, feed the lawn with blood, bone and fishmeal immediately afterwards and mow once a week from April to September.

I haven't used weedkillers for forty years, but I do dig out rosette-forming weeds such as dandelions and plantains with an old tool called a daisy grubber and my lawn looks green and crisp and even as a result. This regime is also kinder to all forms of wildlife. I use a rotary mower with a large rear roller to give me the stripes; keen groundsmen will use a cylinder mower – one of those with blades carried in a helix formation that cut in a scissor action against a lower blade. Cut the grass down to 2.5cm when you feel it needs cutting in the winter months (it will still grow in mild weather) and to half that between April and September (except in dry spells, when raising the height of cut will make the lawn more drought resistant).

NEVER water an established lawn. Yes, the grass will turn straw-like in prolonged dry spells, but you are conserving a valuable resource by not squandering water and the lawn will green up after the first shower of rain. When you are creating a new lawn from seed or turf, then *do* water it in dry spells for the first few weeks until its roots are established. Use a lawn sprinkler left in place for half an hour on any one spot. Only in these circumstances is watering a lawn acceptable, and you'll have to do far less of it if you make a new lawn from seed or turf in September rather than in spring or early summer

# Famous gardener of the month
## David Austin (1926–2018)

Every so often a gardener comes along who changes the very appearance of our gardens, either by their approach to landscape design or by breeding new plants. David Austin devoted his life to breeding roses. Born in Albrighton, Shropshire, it was there in 1969 that he founded the firm that bore his name, breeding and bring-ing to our gardens roses that had the charm and, in most cases, the rich fragrance of old-fashioned shrub roses. They had one vital addition: the ability to flower repeat-edly through the summer, rather than in one brief but glorious flush in late June and early July as is the case with older varieties. His first great success was the single-flush variety 'Constance Spry' in 1961. Eight years later, the first of his repeat-flowering 'English Roses', as they came to be called, encouraged more people to look further than the rather stiff hybrid tea and floribunda roses if they wanted their roses to bloom from May to October. The Austin roses have complex flower forms reminiscent of the old roses along with a vigorous habit and, in many cases, greater disease resistance.

Among David's most popular and finest varieties are 'Graham Thomas', a clear yellow, 'Gertrude Jekyll', rich pink, 'Lady of Shalott', pale orange, 'The Generous Gardener', a wonderful repeat-flowering pink climber, which I cherish on the front of our house. Oh, and he also named a vigorous pink-flowered, peony-shaped rose 'Alan Titchmarsh'. I blush almost as much as it does. David was awarded the Royal Horticultural Society's highest award, the Victoria Medal of Honour, in 2003 and the OBE in 2007. He died in Albrighton, the place where he was born, in 2018 at the age of ninety-two. His glorious legacy lives on.

# Herb gardening

It's not exactly spiritually or gastronomically enriching when you have to resort to that little carousel of cylindrical jars packed with dried-up leaves to spice up your cooking. Yes, it's handy, and with 'mixed herbs' quite effective, but it's so much more enriching to grow your own. What do you need? A spot in full sun and a patch of well-drained soil, and that's about it. There are just a few things to remember: mint is invasive and will take over unless it is planted in a sunken pot or bucket; basil does best in pots indoors on a windowsill, being slightly tender and susceptible to poor weather; and bay will grow into a tree, rosemary into a stout shrub.

Raised wooden planters-cum-troughs sold as miniature veg gardens will raise your herbs to handy picking
height and you'll be able to fit in a surprising selection
– thyme, rosemary, parsley, sage, chives, coriander . . .
whatever you want . . . replacing those that fade with a
new potful from a local nursery or garden centre. Make
your herb garden as close to the kitchen door as you can,
then you won't have to traipse too far in wet weather.
The flavour of those that are fresh far outstrips the dry
dust in those jars.

# Wild flower of the month

## Bluebell (*Hyacinthoides non-scripta*)

Few sights are as breathtaking (or as hard to photograph
– why do they always come out with a pinkish tinge?) as
a counterpane of bluebells on the woodland floor. The
tufts of salad-green leaves appear in February, but the
flowers themselves – arching 30-cm high stems bearing
a nodding row of elfin bells – open in April. They last
for several weeks before fading and producing seed pods
that split when ripe to continue extending the colony.

Dappled shade is what they most enjoy and they are
not to be confused with the altogether beefier Spanish
bluebell (*Hyacinthoides hispanica*), which is most often
to be found in gardens and should on no account be
transplanted into native woodland. (It is a robust brute,
not ugly but by no means as graceful as our native bluebell.) Bulbs can lie dormant for years, as I know from

experience, having cleared a dense plantation of Lawson cypress originally planted as a source of matchwood. The earth below the cypresses – overshadowed and gloomy – was totally devoid of plant life until the trees were felled. The following spring, up came the dormant bluebells – magical! If you want them in your garden, check that the stock you buy is from a cultivated source – they should be protected in the wild.

*Other local common names*: Blue Bonnets, Blue Bottle, Blue Goggles, Crawfeet, Crow-bells, Crowtoes, Cuckoo's Boots, Culverkeys, Fairy Bells, Gooseygander, Grammer-greygles, Griggles, Locks and Keys, Pride of the Wood, Ring o' Bells, Rook's Flower, Snake's Flower, Snap Grass, Wood Bells.

# Things you can do

— Get on with sowings on the veg patch and with hardy annual flowers in beds and borders.
— Mow the lawn once a week from now on, weather permitting.
— Plant border perennials, trees, shrubs and roses from containers.
— Plant gladioli.
— Move evergreen shrubs and plant new ones.
— Start to harden off bedding plants before planting them out next month.
— Prick out seedlings into trays or small pots.
— Move pot plants into larger containers and fresh compost.
— Plant second early and maincrop potatoes and onion sets.
— Pull earth around the emerging shoots of early potatoes to protect them from frost.
— Plant tomatoes in a greenhouse.
— Plant melons and cucumbers in a greenhouse late in the month.
— Make new lawns from seed (and turf, but seed is cheaper).
— Plant globe artichokes and asparagus.
— Sow French and runner beans in pots in a greenhouse.
— Plant cucumbers if your greenhouse is heated.
— Sow aubergines and sweet peppers in a propagator.
— Plant out sweet peas at the foot of bamboo or hazel wigwams.

— Sow parsley.
— Thin out the fruits on peaches and nectarines.
— Pot up begonia tubers in an unheated greenhouse.

# Things you should not do

— Don't buy summer bedding plants unless you can protect them from frost.
— Don't plant out summer bedding yet . . .
— Don't start spraying with insecticides – let nature's predators help you.
— Don't sow seeds of crops your family do not like eating!
— Don't knot the foliage of faded daffodils and narcissi, or tie it up with elastic bands. It needs to flop for six weeks to feed the bulbs – then you can chop it off.

May

As full of spirit as the month of May,
And gorgeous as the Sunne at Midsummer.

WILLIAM SHAKESPEARE, *Henry IV*, *Part 1*, *c.* 1597

In May the shoulders come down, the breathing becomes easier, the sweater comes off and the water-proofs *usually* get a chance to dry out. There is something about May – the very positive nature of its name does a lot to help. If it were called 'Maybe' the feeling would not be the same. But it's not. It's May – and you can. Spring may have officially started five or six weeks ago, but by May the season is fully established and from now on we travel headlong towards summer. More and more flowers open every week.

The flowers on hedgerow hawthorn might have faded by now in an early season, but it gets its other common name – May – from the month in which it normally flowers, when great cataracts of creamy white blossom foam from its cascading branches, releasing their sickly sweet aroma onto the mild air. Other wild flowers keep it company – cow parsley and vetches, foxglove and speedwell, stitchwort and campion – proving that May is a month of wondrous fecundity. On that sunny day, when all the world's vegetation is fresh and green – in garden and countryside – any gardener will feel glad to be alive.

# Weather

Alas, I must offer a word of caution when it comes to assuming that even tender plants will be safe in the garden this month. They *should* be, and maybe they *will*, but if you are wise you will keep an eye on the weather forecast. Frosts have been known to occur in mid-May, and only at the end of the month will you be pretty much guaranteed free of them unless you live in Scotland, in which case you'd best wait until early June before planting out summer bedding. And that is my final word of caution.

Like any other month, May will offer sunshine and showers, grey days and glistening days when the natural world seems to shine, but with any luck the fine days will predominate and the earth will be welcoming to all form of life committed to it. Sow seeds, take cuttings, plant hardy perennials and container-grown shrubs. Fill the gaps in your beds and borders in May and you will enjoy the results in the summer that follows.

# Day length (Skipton)

|  | 1 May | 31 May |
|---|---|---|
| **Dawn** | 04.49 | 03.52 |
| **Sunrise** | 05.29 | 04.41 |

| Sunset | 20.41 | 21.30 |
|---|---|---|
| Dusk | 21.20 | 22.18 |
| Daylight hours | 15 hours 5 minutes | 16 hours 41 minutes |

# Feast days for gardeners

| 1 May | **May Day** – a time for maypole dancing and general rejoicing. |
|---|---|
| **1st Week in May** | **Malvern Flower Show** – in a glorious setting below the Malvern Hills in Worcestershire. |
| **3rd Week in May** | **Chelsea Flower Show** – held in the grounds of the Royal Hospital Chelsea since 1913 and a high spot in the horticultural calendar. |
| **Late May Bank Holiday** | a time to risk planting out summer bedding. |

# Tree of the month

## Snowy mespilus (*Amelanchier lamarckii*)

If you want a modest-sized tree that really earns its keep, choose the snowy mespilus. A native of North America, it has clusters of wonderful starry white blossoms, which cover the tree at the same time as the leaves emerge. When they begin to unfurl they are a dusky pink, and a great foil for the flowers. Over the next few weeks they will turn green, but in autumn the tree is sensational once more when the leaves take on fiery tints of bright red and orange before they fall.

Plant it at the back of a border where it is tall enough to offer stature without casting too much shade, or use it as a focal point in a small garden by planting it at the end of a path. If you want to cheer it up in summer, when it is at its most muted, train a clematis up through the branches so that the flowers can decorate the tree's framework when its springtime spectacle has abated and its autumn glory is some months off.

Height and spread after 20 years: 4m.

# Music to listen to . . .

May is a time for something cheery . . . and cheesy in the case of *Country Gardens*. Essentially British in feel, it was written by Percy Grainger, who was Australian by birth and a great friend of Frederick Delius (see April).

Delius's own composition *La Calinda* is suitably May-like in its appeal, too, along with *The Walk to the Paradise Garden* from his opera *A Village Romeo and Juliet*. You'll find Patrick Doyle's soundtrack for the film *Sense and Sensibility* uplifting and soothing in equal measure, and while we're in a Jane Austen mood (yes, I know we had Dario Marianelli's music for *Pride and Prejudice* in March, but May seems to suit Miss Austen even better), the soundtrack to *Becoming Jane* by Adrian Johnston is perfectly pitched. Those who spurn film soundtracks in favour of more legitimate classical fare can take a holiday and wallow in Mendelssohn's *Symphony No.4* – the *Italian*.

## Vegetables from the garden this month

Asparagus, sprouting broccoli, summer-sown cabbage, lettuce (sown under cover in March), mustard and cress (on a windowsill or in a greenhouse), spring onions, radish, rhubarb.

## Vegetables in store

A few onions might be left, along with potatoes, but from next month you'll be eating fresh produce from the garden. Hurrah!

# Something to muse upon . . .

H.E. Bates is perhaps best known as the creator of The
Larkins, that earthy, good-humoured family whose
adventures he chronicled in *The Darling Buds of May*
and subsequent volumes. But Bates was also an accomplished gardener. I seem to remember that a part of his
garden was planted so that it did not come into bloom
until September, when the gardens of all his friends
were past their best and over the hill. That, he said,
gave him enormous pleasure. In his book *A Love of
Flowers*, he talks about his garden in Kent and has sage
advice for all those who cherish belief that one day their
garden will be finished:

> The garden I shall describe is an unfinished garden. It
> is also not large. Though it was first hacked out of a
> derelict Kentish farmyard nearly forty years ago, it is
> still not only unfinished but it will never be finished.
> This is because I hold that gardens should never be
> finished. The garden that is finished is dead. A garden
> should be in a constant state of fluid change, expansion,
> experiment, adventure; above all it should be an inquisi-
> tive, loving but self-critical journey on the part of its
> owner. It should in fact reflect its owner. This is why
> gardens ordered and devised according to an expert plan
> are, at best, impersonal and at worst as soulless as the
> sky-scraper blocks of 'packet-of-fags-on-end' architec-
> ture that now curse London and other cities.

The true gardener, like a true artist, is never satisfied; his inherent self-criticism constantly prods him on towards better things, fresh discoveries, new adventures. This dissatisfaction, with me, takes the form of seeking to change some part of the garden every year. A path here, a border there, a bed of something else over there – why, I suddenly ask myself, are they there? Why have they been there, like that, fixed in area and shape, for years? When the answer is, 'I'm damned if I know – I suppose I've just let them go on being there,' I begin to know that my garden needs to suffer a sea change.

I have a sneaking admiration, too, for another of his opinions:

I am neither a lover nor a hater of formality for its own sake, just as I am not sold on any kind of over-natural effect. The two can marry well, each enhancing the other. Nor am I an adherent of the one-colour border school, which gives us wholly blue or white borders at Sissinghurst and wholly white ones at Glyndebourne and elsewhere. They, for me, are boring and I am not at all sure that they are not snobby too. There is more than an aura of good taste about them. I get a sneaking impression that they are born of fear.

H.E. BATES, *A Love of Flowers*, 1971

# Flower of the month

## Peony

A friend of mine refuses to grow peonies. 'But they are in flower for such a short season. I don't see the point in growing them,' he says. Oh dear. Nothing I can say appears to convince him to give them a go.

I have tried explaining that planted among other border perennials their foliage is reasonably handsome once the glorious flowers have faded, and that it does not take up too much room. I have suggested that he do as I do and grow a couple of rows of them on his veg patch so that he can cut them for the house and enjoy them in all their glory at close quarters. To no avail. He will have no truck with them. Contrast this with the frequent requests I receive to recommend a plant that will grow to a manageable height and spread, which will need little pruning and will flower all the year round. There is no such thing, and even if there were, the gardener who owned it would be bored rigid by its unchanging physiognomy. After a couple of months it would cease to be noticed at all.

No, we need plants in our gardens that have a brief but utterly glorious season of interest and the peony is unmatched for spectacle in May. From great yam-like roots (which you must not bury too deeply or the plants will be shy of flowering for years), they send up crimson shoots in March. These turn green with maturity and the flower buds sit atop them like fat penny-lollipops (well, that's what they cost when I was a boy).

These buds burst open and transform into massive double flowers, or else single blooms with a prominent central boss of stamens. The flowers may be crimson, pink, salmony orange or white and they have more individual 'wow' factor than almost any other flower in the garden.

Look for varieties like the ludicrously named 'Pillow Talk', which is soft pink, and 'Coral Charm', whose salmon blooms start pink, then turn pale orange and finally yellow before they fade. 'Bowl of Beauty' has an outer row of rich pink petals and an inner ruff of narrow, creamy-white petaloids; 'Sarah Bernhardt' is an old favourite with fully double flowers of pale pink. All of them will grow in a sunny or very gently shaded spot in well-drained soil, acid or alkaline.

Height and spread: around 75cm.


# Flowers in the garden

Aesculus (horse chestnut), Alliums, *Alyssum saxatile*, Amelanchier (snowy mespilus), Anemones, Aquilegia (columbine), Arabis, Arbutus (strawberry tree), Astrantia (Hattie's pincushion), Aubrieta, Azaleas, Baptisia, Bellis (double daisies), Berberis, Brunnera, *Buddleja globosa*, Callistemon (bottle brush flower), Caltha (kingcup), Camassia (quamash), Campanula, Cardamine (lady's smock), Ceanothus (Californian lilac), Centaurea (perennial cornflower), *Cercis siliquastrum* (Judas tree), Cheiranthus (wallflower), Choisya (Mexican orange blossom), *Clematis montana*, *Clematis alpina* and other early-flowering varieties, Convallaria (lily-of-the-valley), Cornus (dogwoods), Corydalis, Cotoneaster, Crataegus (hawthorn), Cypripedium (lady's slipper orchid), Cytisus (broom), Daphne, *Davidia involucrata* (pocket handkerchief tree), Deutzia, Dicentra (bleeding heart), Digitalis (foxglove), Doronicum (leopard's bane), Embothrium, Eremurus (foxtail lily), *Erica mediterranea*, Erythronium (dog's tooth violet), Euphorbia (spurge), Fritillarias, Genista, Gentians, Geums, hardy Geraniums, Halesia, Helianthemum, Hemerocallis (day lily), Hyacinthoides (bluebell), Iberis, Iris species and varieties, Kerria (Jew's mallow), Kolkwitzia (beauty bush), Laburnum, Leptospermum, Lithospermum, Magnolias, Malus (apple and crab apple), Meconopsis (Himalayan blue poppy), Muscari (grape hyacinth), Myosotis (forget-me-not), *Narcissus* 'Old Pheasant Eye', Olearia, Omphalodes, Ornithogalum (star of Bethlehem), Papaver (Oriental poppy), Phlox (creeping varieties), Peonies, Pieris, Polygonatum (Solomon's seal), Potentilla, Primulas, Prunus (flowering cherry), Pyracantha (firethorn), Rhododendrons, *Ribes speciosum* (fuchsia-flowered gooseberry), Robinia (false acacia), *Rosa hugonis* (and some other species roses), Rosmarinus (rosemary), *Rubus* 'Tridel', *Saxifraga* x *urbium* (London pride), Skimmia, Spiraea (bridal wreath), Sophora, Sorbus (mountain ash), Syringa

(lilac), Tamarix (tamarisk), Tiarella, Trilliums,
Trollius (globe flower), Tulips, Ulex (gorse),
Verbascum (mullein), *Veronica gentianoides*,
Viburnums, Vinca (periwinkle), Violas, Wisteria.

# Herbs from the garden

Bay, chives, marjoram, mint, parsley, rosemary, sage,
and a whole host more from now on into the summer.

# The postponement of gratification

The very term is possessed of echoes of want and
longing. Indeed, I first heard it from a monk who was
extolling the virtues of his monastic existence;
explaining that his reward for a life of privation would
come in heaven. It seemed to me such a long time to
wait.

But I have learned, over the years, to enjoy the anticipation involved in making a garden: in planting an
asparagus bed and foregoing a harvest until the patch
has been established for three years. There is practical-
ity here as well as patience. The delay is not simply a
form of horticultural masochism; it allows the plants to
build up their strength, so that they are able to withstand being plundered of their shoots thereafter and
retain sufficient strength in their rootstocks to push up
more spears each year, in spite of being cut off at their

stocking tops between April and June. And, oh! the heightened delight at the moment of harvest.

Peonies flower but briefly in May and June – but the wait for those globular buds to burst open makes the ridiculously spectacular show all the more appreciated, provided you have come to understand . . . the postponement of gratification. 'It is better to travel hopefully than to arrive,' we are told. If we enjoy the mouthwatering anticipation of the cutting of asparagus and the flowering of peonies, how much greater the pleasure of their arrival than if they were available all the year round? Enjoy the wait and appreciate the brevity of the pleasure, which keeps that longing fresh.

You want instant results? Grow radishes.

# A garden to visit in May

April and May are the rhododendron and azalea months, when woodland gardens on acidic soil (not containing chalk or limestone) erupt with colour. Living on chalky soil in East Hampshire, I have to grow rhododendrons, azaleas and camellias in tubs of ericaceous (lime-free) compost, but just down the road from me, where the chalk peters out and the ground is acidic, enthusiasts flock to Exbury, long established by a branch of the Rothschild family as a showcase for these brilliant spring-flowering shrubs.

There are 200 acres of informal woodland in a garden established over 100 years ago. The area known as the

Azalea Bowl is especially stunning – the brightness of the flowers really will take your breath away. From my point of view, this garden offers a double whammy in that it has a steam railway (always had a fondness for them). The upside of this is that if you have a family member who is not remotely interested in gardening, they will be entranced by the narrow gauge railway that runs through the grounds.

Other gardens with good rhododendron and azalea collections include Coleton Fishacre in Devon, Cragside in Northumberland, Glenarn Gardens in Argyll and Bute, The Isabella Plantation in Richmond Park, Savill Gardens in Surrey, The Lost Gardens of Heligan in Cornwall and Mount Stewart in County Down.

# National Garden Scheme

Founded in 1927, the National Garden Scheme is, quite simply, a phenomenon, and one which achieves an enormous amount of good by contributing millions of pounds each year to nursing charities. The premise could not be simpler: private gardens in England and Wales (Scotland has its own equally valuable scheme) are opened to the public by their owners on pre-arranged dates that suit the type of garden and the season at which that garden looks its best. There are tiny cottage garden plots and large estates, village streets with several gardens open at once, and all have been vetted to make sure they have at least 45 minutes of interest to anyone looking round (and the vast majority will take even longer to explore).

In a single year the charity will donate something in the region of £3 million to charities such as Macmillan Cancer Support, Parkinson's UK and Marie Curie, and smaller charities like Perennial (The Gardener's Royal Benevolent Society) and Horatio's Garden, which creates therapeutic gardens at hospital spinal units across the country. Several thousand gardens open each year, from early spring right the way through to autumn, and they are listed county by county in the fat little 'Yellow Book' that is published by the NGS each year.

Many of the gardens offer refreshments, all of them offer friendliness and hospitality – gardeners tend to be a welcoming bunch – and as well as having a great day out, you will be contributing to vital nursing charities that really do make a difference.

# Bird of the month

## Cuckoo (*Cuculus canorus*)

'The cuckoo comes in April. She sings her song in May. In the middle of June she changes her tune. In July she flies away.' One of a variety of rhymes that explain the activities of a grey, 30-cm-long, dove-like bird with a belly barred in black and white. It has become increasingly rare in England, but the familiar song is cherished by those who hear it floating across the meadow or the marshes. Oh, and it is the male, not the female, which sings the classic song. The cuckoo's baby-minding habits are such that we should not be quite so admiring, but its migration from the forests of the Congo to the UK is nothing short of breathtaking.

The cuckoo leaves Africa by night, flying at a height of between 3 and 5km above the ground. It crosses the Sahara Desert in the space of one 50–60-hour flight – three nights and three days of flying non-stop – before it rests and feeds once more. The reverse migration routes from the UK to Africa in late summer vary depending on the bird's location. Dry summers in south-west Europe leading to poor food supplies have adversely affected the survival rate of the English cuckoo population compared with birds from Wales and Scotland. These take a more south-easterly route (through Italy rather than sun-scorched Spain), which has resulted in the decline of these populations being much slower – stable, in fact, in Scotland.

On arrival in the UK from Africa, the female cuckoo will locate the nest of a host species particular to it – a bird such as the dunnock, the reed warbler or the meadow pipit. Each cuckoo knows the species to which it is suited. In such a nest she will eat a single egg and replace it with one of her own (with markings that mirror those of the host species). She has to be quick – the entire operation can be undertaken in ten seconds. Should the host bird find her at her nefarious work, then the imposter egg will be destroyed, as will any whose markings are significantly different from her own. On hatching, the young cuckoo will heave out any other chicks and any eggs that have been laid by the host bird, ensuring that it is the only mouth to feed.

The cuckoo chick, significantly larger than its diminutive host, is fed by the adoptive parent for three weeks before it fledges, and for another couple of weeks before it becomes independent. Insects and grubs form the basis of its diet. It never sees its real parents, but then the female cuckoo is still busy laying a dozen or more eggs in suitable nests across its territory before, at the end of June, it heads back to Africa to be followed a few weeks later by the offspring it has never met.

This story of deception has intrigued countless generations and continues to be a source of fascination to naturalists, who now know more about the birds' migration habits than ever before. Listen out from April onwards: St Tibertius' Day on 14 April is said to be the first day you will hear the cuckoo's distinctive song. Here in my part of Hampshire, its call usually echoes across the meadows and woodland a week or so later around St George's Day, but I suppose much depends on the weather the cuckoo encounters on that long, arduous and remarkable journey from the forests of the Congo.

The UK population amounts to 15,000 breeding pairs. Here's hoping it continues to thrive.

# Asparagus

Tradition (and common sense) means that you will not harvest any asparagus spears until the third spring after planting. The idea is to allow the plants to build up their strength before you begin to plunder their bounty in April and May in the third year of growing. But it is worth the wait. Slice the 15-cm long shoots from their roots with a sharp knife (a serrated asparagus harvester if you want to do things properly), steam them for five minutes and eat them smothered in butter or mayonnaise or whatever takes your fancy. (Yes, you will find you have smelly wee within half an hour of eating asparagus, but we will pass swiftly over that.)

There is much mystery surrounding the cultivation of asparagus, but in reality it is relatively simple. For a start, it needs a patch on the vegetable plot all to itself. It is a perennial and will be happy to grow in the same spot for years. Plant the one-year-old crowns (clusters of roots that radiate outwards from the central dormant buds) in March. Plant them 5cm deep and 30cm apart in rows 45cm apart. Cultivate the soil well beforehand and work in plenty of well-rotted garden compost or manure. The easiest way to plant is to take out a shallow trench and space out the crowns before backfilling with soil.

Asparagus prefers full sun, but will tolerate a little light shade. Water the plants in dry weather. The shoots will emerge and over a few weeks will turn into tall, feathery wands – a veritable forest of foliage. Cut this down to ground level in autumn and give the bed a 5-cm mulch of manure or compost. The same routine should be followed in year two. By the beginning of the third year, the plants will be well established and you can cut the emerging spears as soon as they are long enough to harvest. Cut *all* of them, and carry on cutting until the middle of June, at which point the plants are left to grow unimpeded by the knife. The foliage they produce now will feed the roots and give them enough strength to produce more spears the following year. The harvesting period lasts about ten weeks, after which you will be ready for a change of diet . . . until next year.

When it comes to varieties, choose one which produces all-male plants. These will not produce seeds and subsequent seedlings that are of dubious quality. The old favourite variety is 'Connover's Colossal', but I prefer to grow 'Ariane', which is remarkably tender and flavoursome. 'Ginlim' and 'Pacific 2,000' are also highly recommended. Take your pick . . .

# Fruits in store

A few late apples, plus berries in the freezer and jams.

# Fruit of the month

## Gooseberry

The gooseberry bush may not be the most friendly of garden inhabitants – it is thorny and rather difficult to work around – but its fruits are simply delicious in pies and fools and crumbles and it will grow in any decent soil and a sunny position. What's more, it crops quite early in the season – the earliest of the soft fruits (those that grow on canes and bushes rather than trees) – and if you want to thin out the fruits in May, then they can be used in advance of the fatter, mature crop.

You can find green gooseberries, yellow gooseberries, and those that are wine red, but flavour is everything. I'd plump for 'Early Sulphur' (yellow) if you want the earliest berries, 'Rokula' (red) another early variety, which is mildew resistant, and the old favorite 'Leveller' (green, turning yellow), which still has succulent berries with a cracking flavour. The bushes are grown on a 'leg' – a short, single stem, but you can also buy 'standard' gooseberries on tall stems that are delightful in an ornamental kitchen garden and will save you bending down to pick the fruits. (Our old Labrador used to jump up for the berries. They wreaked havoc with her digestion. I just drop it in . . .).

Pruning is simple – shorten the new season's growth to around finger length in summer, once the crop has been picked. Try to keep the centre of the bush open to improve air circulation and reduce the risk of that disfiguring white powdery mildew. Gooseberry sawfly can reduce the leaves to a strand of central midrib in summer, but keep the bushes well fed and watered and they will recover. Every garden should have a goose-berry bush – and not only so that small children will know where babies come from.

# Something to read . . .

I don't, as a rule, settle down to read a gardening book cover to cover. I dip. I use them for reference and to check a fact or two. There are exceptions to this rule – the books of Christopher Lloyd (see February) are among those in which I can immerse myself – and a few months ago I found another: *Adventures of a Gardener* by Peter Smithers. Sir Peter (1913–2006) was an MP, a diplomat, from 1964–1969 Secretary General to the Council of Europe, and his book is an odyssey of his gardening experiences in the UK, particularly in Winchester but also overseas. At Vico Morcote, the villa he built by Lake Lugano in Switzerland in 1970, not only did he create the most glorious house, with sliding screen doors to frame the matchless views, but he also constructed and planted a garden where he could grow an astonishing range of treasures.

Sir Peter was that great rarity: an observant gardener of sound practical skills and reasoning, who could also set down in the most infectious manner stories of plants found in the wild and the means of cultivating them in a garden. All too often such expositions can die on the page, but in *Adventures of a Gardener* you will find yourself on a journey you are never likely to make in person, and in the company of the most engaging guide. The book is learned, sparky and opinionated, but always relies on persuasion rather than hectoring when it comes to suggest how a plant might be grown.

You will gather from this that I am rather enamoured of Sir Peter and his adventures. You would be right. He is the most delightful companion in print. I am only sorry that I never got to meet him in person.

# Famous gardener of the month

## Beth Chatto (1923–2018)

I did get to meet Beth Chatto. Rather more than that, she became a great and valued friend whose knowledge as a plantswoman has seldom been equalled before or since. But she also made very good soup. And was wonderful company. She taught us all so many things about plants – most importantly abiding by the dictum of 'right plant, right place' – a belief in comparing a plant's natural habitat with the conditions prevailing in a particular part of her garden and getting the two to match up. The foundation of this belief she credited to her husband Andrew, a former fruit grower with a deep and abiding interest in and knowledge of ecology. But Beth also had a way of placing plants alongside one another to great effect.

They were juxtaposed so that they enhanced each other's attributes, and that was an innate skill she owed to no one, except perhaps to the artist Sir Cedric Morris, in whose garden at Benton End she learned much about plants, and art, and people. To spend a day in Beth's company was one of life's greatest pleasures. In her garden at Elmstead Market near Colchester, in Essex, she grew a wide range of plants, many of them hitherto almost impossible to find. She opened a nursery, 'Unusual Plants' – it still thrives – and for many years, until the effort became too great, she exhibited at Chelsea Flower Show, invariably winning a Gold Medal.

For those who did not get to meet her, and who did not find her smiling over the top of her rimless half-moon glasses and melting their hearts, she left behind several invaluable volumes of wisdom, including *The Dry Garden*, *The Damp Garden*, and a book of correspondence with her great friend Christopher Lloyd: *Dear Friend and Gardener*. Pick up any of them and benefit from Beth's hard-won experience and sound common sense.

# Gardening by the moon

Of course, there are those who think the whole thing is bonkers: something to be practised by druids or people with beards and sandals. (I wish to make it clear that I have absolutely nothing against druids, or beards, and my prejudice against sandals is confined to those who wear them with socks.) Others take a more considered approach. The tides are governed by the moon as it waxes (appears larger) or wanes (appears to shrink). The fact that the phases of the moon can hold sway over such enormous bodies of water as the seas and oceans of the world call into question the effect of the phases of the moon on the sap within a plant.

Devotees of 'Lunar Gardening' suggest that it is best to sow and plant crops that grow above ground when there is a waxing moon, since that is when growth is most noticeable and plant energy at its most vibrant and active. When the moon is waning – from the full moon to the third quarter – it is a good time to sow crops that mature underground. Sow nothing on the fourth quarter of a waning moon, when energy will be diminished.

There is usually one full moon each month – twelve in all – except when, due to planetary activity not tying up exactly with our Gregorian calendar, a thirteenth moon appears: a Blue Moon. Hence the saying, 'Once in a blue moon'. Blue moons are infrequent; occurring every two or, more often, every three years. So . . . if you want to see whether the idea of lunar gardening works in practice, check the phases of the moon and sow and plant accordingly.

I have a wristwatch that shows the phases of the moon. I need to check it more regularly when I am sowing and planting on the veg patch.

# Wild flower of the month

## Cow parsley (*Anthriscus sylvestris*)

There is something about the arrival of cow parsley that fills me with joy. It erupts from the hedgerows in the lanes around our village in May and tells me that winter is well behind us and summer is imminent. I love its graceful habit, the frothiness of its flowers held above the angled stems and ferny foliage. And then there is its alternative name, Queen Anne's lace.

Why? There are various apocryphal stories: that Queen Anne suffered from asthma and in the month of May would take the air in the countryside around Kensington (when countryside still existed in London SW10). She and her ladies-in-waiting took with them their small pillows and carried out their lace-making as they sat among the wild flowers. Perhaps the flower inspired their patterns. Or, that when Queen Anne travelled further afield in May, the locals said that the lanes and streets had been decorated for her royal progress. Either way the name has stuck – but not so much as the rather more rustic and less regal sounding cow parsley.

Another alternative name is mother die – said to be the tragic consequence of picking the flowers. Superstitions abound with wild flowers – both May blossom (hawthorn) and bluebells are said to bring bad luck if picked for decorating the home. Cow parsley, though, is more reliable when picked for a vase – lasting around a week – and is occasionally used in flower

arrangements in country churches during May to celebrate nature's bounty and the effulgence of spring.
Hopefully the influence of the Almighty will overcome
any ill luck.

*Other local common names*: Adder's Meat, Bad Man's
Oatmeal, Devil's Oatmeal, Devil's Parsley, Cow
Chervil, Dog Parsley, Gipsy Curtains, Gipsy's
Umbrella, Hare's Parsley, Honiton Lace, Kedlock,
Kesk, Lady's Lace, Moonlight, Rabbit's Meat, Scabby
Hands, Sheep's Parsley, Wild Parsley.

# Things you can do

— Make new ponds and plant aquatics.
— Stake border perennials before they flop.
— Harden off bedding plants and plant them outdoors towards the end of the month if there is no threat of frost.
— Ventilate the greenhouse and put shading material in place.
— Pot up rooted cuttings and pot on plants that have outgrown their containers.
— Rest nerines and freesias by stopping watering and allowing the bulbs to go dormant.
— Make successional sowings on the veg patch – all kinds of seeds can go in now.
— Plant aubergines and sweet peppers in an unheated greenhouse.
— Earth up early potatoes.
— Plant outdoor tomatoes at the very end of the month and remove sideshoots from those growing in a greenhouse. Feed them once a week from now on.
— Train and feed melons and cucumbers in the greenhouse.
— Thin peaches and nectarines so that the fruits are spaced a few centimetres apart.
— Hoe regularly between veg rows to keep down weeds.
— Water newly emerging crops in dry spells, and any newly planted trees, shrubs and perennials.
— Plant dahlia tubers in beds and borders at the beginning of the month.

— Sow hardy annuals outdoors where they are to flower.
— Harvest asparagus regularly.
— Sow wallflowers on the veg patch for next spring's show.
— Sow French and runner beans and sweetcorn outdoors.
— Push straw under strawberry plants to keep the fruits clear of the soil. Net the crop to repel birds.
— Clear away and compost spring bedding plants to make way for their summer replacements.
— Plant out sweet peas at the foot of wigwams of canes.
— Plant up windowboxes and hanging baskets.
— Thin out raspberry canes if they are overcrowded – aim to space them 15cm apart.
— Sit down and enjoy the view.

## Things you should not do

— Don't panic because you can't keep up.
— Don't water the lawn in dry spells – save water for new plants and veg crops that really need it.
— Don't start spraying to control insect pests. Let natural predators help you out and stop being such a control freak. In time a balance will be struck and epidemics will be rare. Just bite the bullet and give it a go.

June

It ceas'd; yet still the sails made on
A pleasant noise till noon,
A noise like of a hidden brook
In the leafy month of June,
That to the sleeping woods all night
Singeth a quiet tune.

SAMUEL TAYLOR COLERIDGE,
*The Rime of the Ancient Mariner*, 1798

Even the ancient mariner, when he wasn't stopping 'one
of three', mused upon the comforting sounds of a
rippling stream in June – a dreamy image that gives
solace to anyone on dry land as well as those who sail
the ocean blue. June at last, then.

I hesitate to suggest that if your garden doesn't look
good in June then you have a problem. Perhaps it is
kinder to say, that *most* gardens look their *best* in June
– the days are longer, the plants grow faster, the leaves
are a freshly minted green, more flowers open every
week, there are generally enough showers to keep
things growing and . . . it's *warmer*! Frosts are a thing
of the past (except in the Highlands of Scotland
where July is the only month that is reliably free of
frost, but then they do have very long daylight hours
in summer, which makes up for the more subdued
temperature).

June is the month of roses and sweet peas, of
fragrant garden pinks and crispy lettuces, gaps disappear in beds and borders as perennials, trees and
shrubs fatten up for the summer. Everywhere there is a
sense that summer is now underway, except, of course,

for those days when a wind gets up and new growth is rattled and new leaves torn. But we try not to think about that, and hope that such travesties are infrequent. At last we can sit outdoors in the evening with a glass. It may not always be warm, but it will, at least, be light until relatively late. Thank goodness for British Summer Time.

# Weather

The month of June is like the little girl 'who had a little curl, right in the middle of her forehead' – when it's good it's very, very good and when it's bad, it's horrid. 'Flaming June' the month is sometimes called – occasionally in despair. Soft and gentle breezes and refreshing showers of rain are to be expected – and even enjoyed – among the longer, sunny days, but when unseasonable strong winds tear at newly unfurled leaves, then the gardener is understandably despondent about the damage. But perhaps this year will be kinder and we can enjoy the sixth month of the year for what it is – the lightest and the brightest of the months, when everything in the garden is lovely and the scent of roses lingers on the air.

# Day length (Skipton)

|  | 1 June | 21 June | 30 June |
|---|---|---|---|
| **Dawn** | 03.51 | 03.41 | 03.46 |
| **Sunrise** | 04.40 | 04.33 | 04.37 |
| **Sunset** | 21.31 | 21.46 | 21.45 |
| **Dusk** | 22.20 | 22.38 | 22.37 |
| **Daylight hours** | 16 hours 43 minutes | 17 hours 5 minutes | 17 hours exactly |

**21 June is the longest day of the year**

# Feast days for gardeners

| | |
|---|---|
| **1 June** | the start of **Meteorological Summer.** |
| **Derby Day** | the traditional day for clipping box (but wait if the day is scorchingly hot, lest the cut leaves scorch in the sun). |
| **3rd Sunday in June** | **Father's Day** – an American tradition but one we seem to have adopted in the UK. A chance to treat your dad to a new shed or a packet of seeds depending on the available budget. |

**21 June**   **The Summer Solstice** – the start of the astronomical summer and the day with the longest daylight hours (actually, there's nothing in it between this day and the days either side).

**24 June**   **Midsummer's Day** – probably the one Shakespeare was thinking about. He wrote *A Midsummer Night's Dream* around 1594, just twelve years after the Julian calendar was replaced by the Gregorian calendar we use today. The Gregorian calendar more accurately reflects the time it takes the Earth to circle once around the Sun. The old date for Midsummer's Day has been retained.

# Tree of the month

## Laburnum (Golden rain tree)

Few early summer sights are as refreshingly uplifting as that of a laburnum tree in flower. A tree of grace, which drips with its golden chains of pea flowers in June, its season of glory may be relatively brief, but it is certainly memorable. Plant it in the centre or at the back of a bed or border where it can be clearly seen when it is garlanded with blossoms, but where it will also act as a plain background showing off brighter, later blooms around it when its own moment of magnificence passes. Young laburnum trees planted at either side of a pergola – a series of linked arches – and trained over the top, will form a tunnel of floral sunshine in early summer and a shady grotto for the rest of the year.

If you are concerned about small children eating the seeds, which are toxic like all parts of this plant, grow the variety *Laburnum* x *watereri* 'Vossii', which, apart from having the longest and most spectacular flower trails, produces fewer seeds than the common laburnum (*L. anagyroides*). The seeds are small and black and unlikely to be regarded as tempting food by most children, and planted at the back or the middle of a border, they will be unlikely to wander beneath it. The tree needs full sun to flower at its glorious best and it will grow to only a modest size.

Height and spread: 5–6 m.

## Music to listen to . . .

Something relaxed in its tempo seems to suit June, and I know of no piece of music more appropriate than *The Banks of Green Willow* by George Butterworth, which will transport you to the limpid waters of a gin-clear chalkstream or a river in the Yorkshire Dales. Butterworth lost his life at the age of thirty-one in the First World War, and this piece of music is a haunting legacy that makes one wonder what other riches he would have shared had his life not been cut tragically short. There is a wistfulness about this piece of music, but not a sadness, since it simply celebrates the glories of the British countryside he fought to protect.

Another character whose life was devoted to the British landscape was Beatrix Potter who, as well as writing children's books, bought up and conserved parts of the Lake District for future generations – handing them to the National Trust on her death. She also championed the Herdwick Sheep that grazed its fells, making sure that the breed survived and prospered. Remember her when you listen to the soundtrack of *Miss Potter* by Nigel Westlake, as it transports you to the fells of Cumberland and Westmorland (in the days before the two counties became designated as Cumbria in 1974). The music is charming, and Katie Melua's rendition of 'When You Taught Me How to Dance' is hauntingly beautiful.

# Vegetables from the garden this month

Asparagus (but stop cutting in the middle of the month), broad beans, young carrots, lettuce, mustard and cress on a windowsill or in a greenhouse, spring onions, overwintering onions, early potatoes, radish, rhubarb, spinach, turnips.

# Something to muse upon . . .

If I can cajole you into doing one thing, thanks to your acquisition of this almanac, it need not be to make some renewed foray in controlling weeds or diminishing your slug population (though both are necessary aims), it should be to take up a book by the finest humorous writer of his – and almost any other – generation: Pelham Grenville Wodehouse.

His Jeeves and Wooster novels are famed abroad, but my own predilection is for those volumes concerned with the residents of Blandings Castle in Shropshire, viz: Clarence, 9th Earl of Emsworth, his sister Lady Constance Keble, his brother The Hon. Galahad Threepwood, his butler Beach, and the rest of the denizens of that fabled pile. You will be transported, courtesy of the most felicitous style of writing, to an England of yesteryear; to the English countryside and the English garden where the sun shone, the bees buzzed and . . . well, the opening words of *The Crime Wave at Blandings* will put you in the mood:

The day on which Lawlessness reared its ugly head at Blandings Castle was one of singular beauty. The sun shone down from a sky of cornflower blue, and what one would really like would be to describe in leisurely detail the ancient battlements, the smooth green lawns, the rolling parkland, the majestic trees, the well-bred bees and the gentlemanly birds on which it shone.

But those who read thrillers are an impatient race. They chafe at scenic rhapsodies and want to get on with the rough stuff. When, they ask, did the dirty work start? Who were mixed up in it? Was there blood, and, if so, how much? And – most particularly – where was everybody and what was everybody doing at whatever time it was? The chronicler who wishes to grip must supply this information at the earliest possible moment.

The wave of crime, then, which was to rock one of Shropshire's stateliest homes to its foundations broke out towards the middle of a fine summer afternoon, and the persons involved in it were disposed as follows:

Clarence, ninth Earl of Emsworth, the castle's owner and overlord, was down in the potting-shed, in conference with Angus McAllister, his head gardener, on the subject of sweet peas.

His sister, Lady Constance, was strolling on the terrace with a swarthy young man in spectacles, whose name was Rupert Baxter, and who had at one time been Lord Emsworth's private secretary.

Beach, the butler, was in a deck-chair outside the back premises of the house, smoking a cigar and reading Chapter Sixteen of *The Man With The Missing Toe*.

George, Lord Emsworth's grandson, was prowling
through the shrubbery with the airgun which was his
constant companion.

Jane, his lordship's niece, was in the summerhouse
by the lake.

And the sun shone serenely down – on, as we say,
the lawns, the battlements, the trees, the bees, the
best type of bird and the rolling parkland.

Presently Lord Emsworth left the potting-shed and
started to wander towards the house. He had never felt
happier. All day his mood had been one of perfect
contentment and tranquility . . .

P.G. WODEHOUSE, *The Crime Wave at Blandings*, 1937

And if that doesn't tempt you to read on . . .

# Flower of the month

## Delphinium

It is hard not to use superlatives for every flower I
mention, but nothing, I repeat *nothing*, eclipses the
towering grandeur of the delphinium. When the famous
firm of Blackmore and Langdon exhibit them at
Chelsea, crowds gawp in admiration at the 2.4-m spires
of blue, purple or white flowers. There are pinks and
– rarely – reds, too, but it is the heavenly blues I adore,
the colour of a summer sky or rich, deep royal blue,

matched only by a gentian and that diminutive summer annual *Phacelia campanularia*. Grow them at the back of a border where they can be cut back after flowering and the plants in front of them are allowed to take on the decorative mantle.

They need sun, rich, well-drained soil, and protection from slugs and snails when their shoots are emerging from the ground in spring. Sharp grit is a reasonable deterrent; copper collars placed around each clump are more effective. They'll need staking as the stems extend. I tend to loop them into a group with soft twine, corralled between three or four canes so that when the wind does catch them they are allowed some degree of lateral movement. Tie them in individually and they can snap off in strong winds at the point where the tie meets the stem. The clumps can be dug up, divided and replanted in early spring every third or fourth year before they start to lose vigour. If I could have only one variety it would be 'Pandora', whose blooms are an electric blue with a black eye. Just breathtaking.

# Flowers in the garden

Abelia, Abutilon, Achillea, Aconitum (monkshood), Acorus, Aethionema, Ageratum, Alisma, Allium, Alyssum, Anchusa, Andromeda, Anemone, Anthemis, Antirrhinum (snapdragon), Aquilegia (columbine), Armeria, Aruncus, Asperula, Aster, Astilbe, Astrantia (Hattie's pincushion), Azalea, Borago (borage),

Buddleja, Butomus (flowering rush), Calamintha (cala-
mint), Camassia (quamash), Campanula (bellflower),
Carpenteria, Ceanothus (Californian lilac), Centaurea
(cornflower), Centranthus (valerian), Cerastium (snow
in summer), Cistus (sun rose), Clematis, Colutea,
Convallaria (lily-of-the-valley), Coreopsis (tickseed),
Cornus (dogwood), Corydalis, Crambe, Crinodendron,
Cypripedium (lady's slipper orchid), Cytisus (broom),
Daboecia, Delphinium (larkspur and perennial kinds),
Dianthus (pink), Dicentra (bleeding hart), Dictamnus
(burning bush), Digitalis (foxglove), Dryas (mountain
avens), Embothrium, Eremurus (foxtail lily), Erigeron
(fleabane), Erodium, Filipendula, Fuchsia, Gaillardia,
Galega (goat's rue), Gaultheria, Genista, Gaura, hardy
Geraniums, Geum, Gladiolus, Gypsophila, Hebe,
Helianthemum (rock rose), Helichrysum, Hemerocallis
(day lily), Heuchera, Hosta, Hottonia, *Hydrangea
aspera*, *Hydrangea petiolaris*, Incarvillea, Indigofera,
Irises, *Jasminum humile*, *Jasminum officinale*, Kalmia,
Kniphofia (red-hot poker), Kolkwitzia (beauty bush),
Laburnum, Leontopodium (edelweiss),
Leptospermum, Leucothoe, Lewisia, Lilium (lilies),
Linaria (toadflax), Linum (flax), Lithospermum,
Lobelia, Lonicera (honeysuckle), Lupinus (lupin),
Lychnis (catchfly), Lysimachia, Maclaeya, Magnolia
(late-flowering species), Malva (mallow), Meconopsis,
Menyanthes (bog bean), Nepeta (catmint),
Nomocharis, Nymphoides, Oenothera (evening prim-
rose), Olearia, Orchis, Ornithogalum (star of
Bethlehem), Oxalis, Paeonia (peony), Papaver (poppy),

Passiflora (passion flower), Pelargoniums, Philadelphus (syringa), Phlomis, Alpine Phlox, Polemonium (Jacob's ladder), Polygonatum (Solomon's seal), Potentilla, Primula, Pyracantha (firethorn), Pyrethrum, Ramonda, Ranunculus (buttercup), Rhododendron, Rhodohypoxis, Roscoea, Roses, Rubus, Salvia, Saponaria (soapwort), *Saxifraga* x *urbium* (London pride), *Scilla peruvianum*, Silene (catchfly), Sisyrinchium, Smilacina (false spikenard), Sophora, *Sorbus aria* (whitebeam), Spiraea, Stachys, Syringa (lilac), Thalictrum, Thymus (thyme), Trollius (globe flower), Tropaeolum (nasturtium), Verbascum (mullein), Veronica, Vinca (periwinkle), Viola, Weigela, Wisteria, *Yucca whipplei*.

# A garden to visit in June

It's hard to know what to choose, since any garden worth its salt will be a picture this month. That said, this is the month of roses and Mottisfont Abbey in Hampshire is famed for its collection of old and shrub roses. Other gardens that major on the nation's favourite flower include Sudeley Castle in Gloucestershire, and the astonishingly elaborate gardens surrounding what appears to be a French chateau at Waddesdon Manor, near Aylesbury in Buckinghamshire, created by Ferdinand de Rothschild at the end of the nineteenth century. Kiftsgate Court in Gloucestershire is home to the

'Kiftsgate' rose, and Peter Beales Roses at Attleborough in Norfolk, founded by one of the greatest rosarians of the twentieth century, is famed for its collection of shrub roses, many of which are on sale in the nursery. Visit the last named with an estate car to accommodate your plunder.

# Shrub roses

So what is it about shrub roses that makes them so desirable? Well, they are a different animal to the bush roses – those rather stiffly branched hybrid teas and floribundas that filled the rose beds across Britain in the first half of the twentieth century. Shrub roses have a more relaxed habit; many of them grow quite tall and wide. Some of them are species roses, collected from the wild over the centuries, others, like the candy-striped *Rosa gallica* 'Versicolor', known as 'Rosa Mundi', date back to the thirteenth century, and yet more to the centuries that followed. Their histories are as interesting as their flowers; it's no wonder folk get hooked. Then there are cultivated varieties that plant breeders like Peter Beales and David Austin produced in the twentieth century for our delectation and delight.

Shrub roses have many different kinds of flower – from those that are single with a central coronet of stamens, often followed by colourful hips – and fully double kinds that may have what's called a 'button eye', or flowers that appear to have been sliced across with a sharp knife (they look much better than they sound!). Varieties like 'Charles de Mills' (1790), rich burgundy red, and 'Jacques Cartier' (1868), soft pink, have a fragrance to die for.

Most of them are exquisitely perfumed, but their one drawback is that many of them have one brief but glorious season of flowering in late June and early July ('Jacques Cartier' is one of the exceptions to the rule). That was until breeders like David and Peter created shrub roses that were repeat-flowering, blooming off and on from June until late summer, while retaining the handsome flower form and often the fragrance of their forbears.

Pruning is not nearly so severe as it is with bush roses, which are chopped back to around knee height. Shrub roses may have a few older stems removed each year to keep them shapely and youthful, and their stems cut back by about a quarter of their length, retaining the overall shape of the bush. Other than that they are relatively maintenance free, though they best enjoy good soil that is not prone to drying out and a position in full sun. Collect a few of them and enjoy them at close quarters when you pick a bunch of blooms for the house. You can drown in their perfume . . .

# Bird of the month

## Wren (*Troglodytes troglodytes*)

I love the Jenny Wren's Latin name, which implies that it is a cave dweller. Perhaps the fact that it enjoys crevices in which to roost – as well as thickets of ivy – is responsible for that. For me it is not only the wren's neat and agile demeanour – with that needle-like beak and cocked tail – that make it endearing, but the glorious trill coming from one of our most diminutive birds, which, it is said, can carry half a mile. What's more it can be heard at any time of year – not for the wren an 'off season'.

Second only to the robin in its friendliness in the garden, people of a certain age will remember its image on the back of a farthing (a quarter of an old penny), until it ceased to be legal tender in 1960. Its image was used simply because it was the smallest common British bird (the powers that be must have discounted the goldcrest and the firecrest) and the farthing was the smallest value coin. Although it is an LBJ (Little Brown Job), its size and its cocked tail make it easily distinguishable from the likes of the hedge sparrow or dunnock (another tuneful songster). Its busy activity is driven by its need to feed on insects – some of which may be fast moving.

The male builds a selection of egg-shaped nests of moss, feathers, bits of wool and whatever 'fabric' it can find, and in April or May the female, when she has given her approval to her favoured construction, lays one brood of between five and eight white eggs that are speckled with red. Incubation takes 14–17 days and two to three weeks after hatching the young are fully fledged. In winter the birds roost in large numbers – presumably to keep warm. Thirty wrens have been recorded roosting in a house martin's nest, and one winter in Norfolk, sixty birds were recorded in a nest-box measuring 14 x 11.4 x 14.6cm. Now that's snug!

UK population: 10–11 million breeding pairs, making it our commonest breeding bird.

# Staking

In an ideal world our garden plants would support themselves, but modern varieties have often been bred with beauty rather than brawn in mind, and unless they are given some artificial means of remaining upright, they will topple and the display will be ruined. The most natural way of holding up border perennials is to use twiggy pea-sticks, though these are becoming increasingly difficult to find. Push them in among the stems to be supported *before* they are needed, so that the plants can continue to grow and mask their supports. Alternatively bamboo canes and soft garden twine can be used, either as individual supports, or by creating a circle of three or four and looping twine around them to make a loose cage in which the plants can be corralled.

Peonies and the taller sedums (which tend to flop once their flowers open) can be encouraged to grow through circular wire frames on legs, which act as a kind of floral corset. *Hydrangea* 'Annabelle' is a beautiful white-flowered shrub, but her mop-headed blooms are just too heavy for their stalks and will flop as they begin to increase in size. I got our local blacksmith to make me some 1.25-m tall 'coronets' from iron rods, whose circle of uprights (linked by two lower hoops) are bent outwards at the top and finished with a spherical nob – for aesthetic as well as safety reasons. They stay in position all year round and are good looking enough in winter to be admired as garden sculpture.

Sweet peas need tall supports – wigwams of five or six 2m-long bamboo canes or hazel bean-poles, and the same is true of climbing French and runner beans. All of them can also be grown on a tent-like row of these supports on the veg patch. The most important thing is to erect the supports before they are needed, otherwise you'll have a dickens of a job to prevent your plants looking as though they have been trussed up against their will.

# Fruit of the month

## Strawberry

Wimbledon does not have the monopoly on strawberries, even if they do get through 28,000 kilograms of them at each tournament. The great thing about them is that they can be fitted into the smallest space and, when home grown, you can choose varieties renowned for their flavour rather than simply for their appearance and keeping qualities, as is the case with those sold in shops and supermarkets.

My own preference is for the quite delicious 'Sweetheart', but older gardeners (if such there be . . .) will still swear by 'Royal Sovereign', though it can be tricky to grow. 'Cambridge Favourite' lives up to its name, too, and 'Honeoye' is also delicious. The small-fruited alpine strawberries are very tasty, but fiddly to harvest and they don't give you much of a mouthful. That said, it's always worth tasting a strawberry variety before you commit to planting a whole bed of it.

Spring is the best time to plant, even though you might not have much of a harvest in the first year. At least it means the plants can establish themselves in favourable weather. Plant them 30cm apart in rows 45cm apart and remove all runners that form during the growing season – those long, questing shoots with a plantlet at the end. After three years the bed will benefit from being replaced – you can use the plantlets for your new plantation, unless the plants are showing signs of virus disease (mottling and distortion of the leaves). Good soil and a sunny position will produce the best crops, and the plants live up to their name as the fruits form when straw (or black felt mats) can be pushed beneath the fruits to prevent them from being splashed by mud or eaten by slugs.

Birds love strawberries, but if you cover them with nets, do make sure the structure is impenetrable. Blackbirds can become entangled in loose netting and, at the very least, become distressed or even injured. A sturdy fruit cage will make harvesting all soft fruits so much less of a hassle. Pricey? Yes, but long-lasting and well worth the investment.

## Fruit from the garden

Cherries, gooseberries, raspberries, strawberries.

## Something to read . . .

You will have noticed, perhaps, a sprinkling of children's books throughout the almanac. I make no apology for those I have included: they are worth picking up again in adulthood and reminding ourselves of why we were so entranced in the first place.

I have loved *The Wind in the Willows* by Kenneth Grahame since I was a child, mainly for its evocation of life on the riverbank and the depths of the wildwood, but also for its delightful characters. We all know a 'Moley' – shy and retiring, afraid of his own shadow, but more eager than he would admit to experience the adventures he comes to enjoy with the commonsensical Ratty, never happier than when he is 'messing about in boats'. Toad is the egotistical, over-enthusiastic and bombastic extrovert who never seems to notice that he is heading for disaster, and Badger the curmudgeonly stay-at-home who proves a valued friend when push comes to shove.

Then there are the 'Nasty Bits of Work' – the weasels and stoats, who skulk in the shadows and eventually invade Toad Hall. I re-read the book almost every year and notice more readily than I did when I was a child the undercurrents – most especially in the chapter entitled 'The Piper at the Gates of Dawn', with its intimations of a life beyond that lived on the riverbank. Pick it up again and lose yourself among the reeds and the willow wands, the setts and the burrows of our old friends.

# Famous gardener of the month

## Sir Joseph Paxton (1803–1865)

Paxton was a phenomenon, nothing less. Born in Bedfordshire, he began life as a garden boy, eventually finding himself at the Horticultural Society's Gardens in Chiswick, next door to Chiswick House, whose owner, the 6th Duke of Devonshire, spotted his talents and offered him the post of Head Gardener at Chatsworth at the age of just twenty. It was there that Paxton made his name, not only as a fine cultivator of plants, but by masterminding the construction of the Rock Garden, the Emperor Fountain, and the Great Conservatory, designed by Decimus Burton. The structure whetted Paxton's appetite for architectural endeavours and in 1850 he himself designed the structure that came to be known as the Crystal Palace, home of the Great Exhibition, which earned him his knighthood.

How Paxton fitted so much into his relatively short life (he was only sixty-one when he died) is hard to comprehend. He was a director of the Midland Railway (the main source of the wealth he accumulated), Liberal MP for Coventry, designer of railway architecture, organiser of plant-collecting expeditions to North America, a writer and publisher of books and magazines such as the *Horticultural Register* and the *Magazine of Botany*, even co-founding the *Gardeners' Chronicle* in 1841. He designed buildings and houses, including Mentmore Towers in Buckinghamshire, and even commanded the 11th Derbyshire Rifle Volunteer Corps.

On his arrival at Chatsworth on 9 May 1826, he went to breakfast with Mrs Gregory and fell in love with her niece, Sarah Bown. He married her a year later. His story is one of astonishing endeavours and remarkable tenacity. What a man!

# Wild flower of the month

## Dog rose (*Rosa canina*)

There is something delightfully refreshing and simple about the flower of the dog rose. It carries its flowers in clusters and each one has five evenly spaced and gently indented petals of deep or soft pink, sometimes white, arranged around a central coronet of golden stamens.

It is a vigorous shrub that pushes its viciously thorny stems – anything up to 10m long – through hedgerows countrywide, decorating them with its floral confetti in the month of June and July and following the display with a bounty of fat, red hips. As a child, we used to collect these by the bagful and take them into school so that they could be sent off to make health-giving rose-hip syrup, rich in vitamin C, which was later administered to us as a tonic. *Rosa rubiginosa* – the sweet briar or Shakespeare's eglantine – is even more delightful, with foliage that reveals the fragrance of apples when crushed.

*Other local common names*: Briar-rose, Brimmle, Canker-rose, Cat-rose, Cock-bramble, Hip-tree, Dog's Briar.

The fruits are variously known as: Puckies, Cat-jugs, Dog Berries, Hedgy-pedgies, Buckie-berries, Dog-jumps, Canker-berries, Hipsons, Huggans, Nippernails. Children would put the seeds down the necks of their school friends to make them itch, when they were variously known as: Buckie Lice, Cow Itchies, Ticklers and Tickling Tommies.

# British butterflies

'Animated flowers', the Prince of Wales calls them and, indeed, butterflies add mobile colour to our gardens throughout the summer. In a good year at any rate. The thing about butterflies is that they are great barometers, not only of the weather but of wider climate change. The adults may overwinter in sheds and garages, cracks and crevices, and emerge in spring to lay their eggs on plants that will feed their caterpillars – many different species, from grasses to trees and shrubs. Other butter-flies – such as the Clouded Yellow – will migrate from Europe and arrive on our shores in summer to lay their eggs.

Cold, wet and windy weather can adversely affect the survival rate of all butterflies, as can the availability of food supplies, so it is vital that gardeners provide not only nectar-rich flowers (those that are single rather than double) for the adults, but egg-laying sites that will provide food for hungry caterpillars. Many different grasses, vetches, clover, nettles, buckthorn . . . the list goes on, with each species having its favoured nursery. Oh, and the large and small whites will be delighted to lay their eggs on your nasturtiums and cabbage plants. Sorry.

There are around sixty different species of butterfly regarded as being native or frequently encountered in the British Isles. I list them along with the food plants of their caterpillars upon which their eggs are laid:

**Adonis Blue** – Horseshoe Vetch (*Hippocrepis comosa*)

**Black Hairstreak** – Blackthorn (*Prunus spinosa*) and Wild Plum (*P. domestica*)

**Brimstone** – Buckthorn (*Rhamnus cathartica*), Alder Buckthorn (*Frangula alnus*)

**Brown Argus** – Common Rock Rose (*Helianthemum nummularium*), Cranesbill (*Erodium* and *Geranium* spp)

**Brown Hairstreak** – Blackthorn (*Prunus spinosa*) and Wild Plum (*P. domestica*)

**Chalk Hill Blue** – Horseshoe Vetch (*Hippocrepis comosa*)

**Chequered Skipper** – In England: Wood Small-reed (*Calamagrostis epigejos*) and False Brome (*Brachypodium sylvaticum*). In Scotland: Purple Moor-grass (*Molinia caerulea*)

**Clouded Yellow** – Clovers (*Trifolium* spp.), Lucerne (*Medicago sativa*), Common Bird's-foot Trefoil (*Lotus corniculatus*)

**Comma** – Stinging Nettle (*Urtica dioica*), Hop (*Humulus lupulus*), Elms (*Ulmus* spp), Currants (*Ribes* spp), Willows (*Salix* spp)

**Common Blue** – Common Bird's-foot Trefoil (*Lotus corniculatus*), Greater Bird's-foot Trefoil (*L. pedunculatus*), Black Medick (*Medicago lupulina*), Common Restharrow (*Ononis repens*), White Clover (*Trifolium repens*), Lesser Trefoil (*T. dubium*)

**Cryptic Wood White** (only in Ireland) – Meadow Vetchling (*Lathyrus pratensis*), Tufted Vetch (*Vicia cracca*), Common Bird's-foot Trefoil (*Lotus* *corniculatus*), Greater Bird's-foot Trefoil (*L. pedunculatus*)

**Dark Green Fritillary** – Dog-violet (*Viola riviniana*), Hairy Violet (*V. hirta*), Marsh Violet (*V. palustris*)

**Dingy Skipper** – Common Bird's-foot Trefoil (*Lotus corniculatus*), Horseshoe Vetch (*Hippocrepis comosa*), Greater Bird's-foot Trefoil (*Lotus pedunculatus*)

**Duke of Burgundy** – Cowslip (*Primula veris*), Primrose (*P. vulgaris*) and their hybrid

**Essex Skipper** – Mainly Cock's-foot Grass (*Dactylis glomerata*), but also Creeping Soft Grass (*Holcus mollis*), Couch Grass (*Elytrigia repens*), Timothy (*Phleum pretense*), Meadow Foxtail (*Alopecurus pratensis*), False Brome (*Brachypodium sylvaticum*), Tor-grass (*B. pinnatum*) and, rarely, Yorkshire Fog (*Holcus lanatus*)

**Gatekeeper** – Fine grasses such as Bents (*Agrostis* spp.), Fescues (*Festuca* spp.), Meadow Grasses (*Poa* spp.), also Couch Grass (*Elytrigia repens*)

**Glanville Fritillary** – Ribwort Plantain (*Plantago lanceolata*) and sometimes Buck's-horn Plantain (*P. coronopus*)

**Grayling** – Sheep's Fescue (*Festuca ovina*), Red Fescue (*F. rubra*), Bristle Bent (*Agrostis curtisii*), Early Hair-grass (*Aira praecox*) and occasionally Tufted Hair-grass (*Deschampsia caespitosa*), and Marram Grass (*Ammophila arenaria*)

**Green Hairstreak** – Common Rock-rose (*Helianthemum nummularium*), Common Bird's-foot Trefoil (*Lotus corniculatus*), Gorse (*Ulex europaeus*),

Broom (*Cytisus scoparius*), Dyer's Greenweed (*Genista tinctoria*). On Scottish moorland: Bilberry (*Vaccinium myrtillus*). Occasionally: Dogwood (*Cornus sanguinea*), Buckthorn (*Rhamnus cathartica*), Cross-leaved Heath (*Erica tetralix*), Bramble (*Rubus fruticosus*)

**Green-veined White** – Garlic Mustard (*Alliaria petiolata*), Cuckoo Flower (*Cardamine pratensis*), Hedge Mustard (*Sisymbrium officinale*), Watercress (*Rorippa nasturtium-aquaticum*), Charlock (*Sinapis arvensis*), Large Bittercress (*Cardamine amara*), Wild Cabbage (*Brassica oleracea*), Wild Radish (*Raphanus raphanistrum*). Occasionally Nasturtium (*Tropaeolum majus*) and cultivated brassicas

**Grizzled Skipper** – Agrimony (*Agrimonia eupatoria*), Creeping Cinquefoil (*Potentilla reptans*), Wild Strawberry (*Fragaria vesca*). Occasionally: Barren Strawberry (*Potentilla sterilis*), Tormentil (*P. erecta*), Salad Burnet (*Sanguisorba minor*), Bramble (*Rubus fruticosus*), Dog Rose (*Rosa canina*), Wood Avens (*Geum urbanum*)

**Heath Fritillary** – Common Cow-wheat (*Melampyrum pretense*), Ribwort Plantain (*Plantago lanceolata*), Germander Speedwell (*Veronica chamaedrys*). Occasionally: Other Speedwells (*Veronica* spp.), Foxglove (*Digitalis purpurea*)

**High Brown Fritillary** – Dog-violet (*Viola riviniana*), Hairy Violet (*V. hirta*). Occasionally: Heath Dog-violet (*V. canina*) and Pale Dog-violet (*V. lactea*)

**Holly Blue** – In spring: Holly (*Ilex aquifolium*). In summer: Ivy (*Hedera helix*). Occasionally: Spindle

Tree (*Euonymus europaeus*), Dogwoods (*Cornus* spp.), Snowberries (*Symphoricarpos* spp.), Gorses (*Ulex* spp.), Bramble (*Rubus fruticosus*)

**Large Blue** – Wild Thyme (*Thymus polytrichus*) and then Ant Grubs (*Myrmica* spp.)

**Large Heath** – Mainly Hare's-tail Cottongrass (*Eriophorum vaginatum*). Also, Common Cottongrass (*E. angustifolium*), Jointed Rush (*Juncus articulatus*)

**Large Skipper** – Cock's-foot Grass (*Dactylis glomerata*), Purple Moor Grass (*Molinia caerulea*), False Brome (*Brachypodium sylvaticum*) and sometimes Tor-grass (*B. pinnatum*) and Wood Small-reed (*Calamagrostis epigejos*)

**Large Tortoiseshell** – Elms (*Ulmus* spp.), Aspen (*Populus tremula*), Birch (*Betula* spp.), Poplars (*Populus* spp.) and Willows (*Salix* spp.)

**Large White** – *Cruciferae* family (Brassicas, including Cabbages and Brussels Sprouts), Nasturtium (*Tropaeolum majus*), Wild Mignonette (*Reseda lutea*), Sea Kale (*Crambe maritima*)

**Lulworth Skipper** – Tor-grass (*Brachypodium pinnatum*)

**Marbled White** – Mainly Red Fescue (*Festuca rubra*). Also: Sheep's Fescue (*F. ovina*), Yorkshire Fog (*Holcus lanatus*), Tor-grass (*Brachypodium pinnatum*)

**Marsh Fritillary** – Devil's-bit Scabious (*Succisa pratensis*). On chalk grassland: Field Scabious (*Knautia arvensis*), Small Scabious (*Scabiosa columbaria*)

**Meadow Brown** – Many grasses, especially fine-leaved Bents (*Agrostis* spp.), Fescues (*Festuca* spp.) and Meadow Grasses (*Poa* spp.)

**Monarch** – Milkweeds (*Asclepias* spp.), but the butterfly does not breed here

**Mountain Ringlet** – Mat-grass (*Nardus stricta*) and possibly Sheep's Fescue (*Festuca ovina*)

**Northern Brown Argus** – Common Rock-rose (*Helianthemum nummularium*)

**Orange-tip** – *Cruciferae* family, especially Cuckoo Flower (*Cardamine pratensis*) and Garlic Mustard (*Alliaria petiolata*). In gardens: Honesty (*Lunaria annua*) and Dame's Violet (*Hesperis matronalis*) though the larval survival rate is reduced

**Painted Lady** – Wide range of food plants, but especially Thistles (*Cirsium* spp. and *Carduus* spp.) Also found on Mallows (*Malva* spp.), Stinging Nettle (*Urtica dioica*) and Viper's Bugloss (*Echium vulgare*)

**Peacock** – Mainly Stinging Nettle (*Urtica dioica*), but also Small Nettle (*U. urens*) and Hop (*Humulus lupulus*)

**Pearl-bordered Fritillary** – Mainly Dog-violet (*Viola riviniana*), but also Heath Dog-violet (*V. canina*) and Marsh Violet (*V. palustris*)

**Purple Emperor** – Mainly Goat Willow (*Salix caprea*). Also Grey Willow (*S. cinerea*), and Crack Willow (*S. fragilis*)

**Purple Hairstreak** – Sessile Oak (*Quercus petraea*), Pedunculate Oak (*Q. robur*), Turkey Oak (*Q. cerris*) and occasionally Evergreen Oak (*Q. ilex*)

**Red Admiral** – Mainly Stinging Nettle (*Urtica dioica*). Also Small Nettle (*U. urens*),

Pellitory-of-the-wall (*Parietaria judaica*), Hop
(*Humulus lupulus*)

**Ringlet** – Grasses including Cock's-foot Grass (*Dactylis glomerata*), False Brome (*Brachypodium sylvaticum*), Tufted Hair-grass (*Deschampsia caespitosa*), Couch Grass (*Elytrigia repens*) and Meadow Grasses (*Poa* spp.)

**Scotch Argus** – In Scotland: Purple Moor-grass (*Molinia caerulea*). In northern England: Blue Moor-grass (*Sesleria caerulea*)

**Silver-spotted Skipper** – Sheep's Fescue (*Festuca ovina*)

**Silver-studded blue** – Wide range of *Ericaceae* and *Leguminosae* families, especially Heather (*Calluna vulgaris*), Bell Heather (*Erica cinerea*), Cross-leaved Heath (*E. tetralix*), Gorses (*Ulex* spp.)

**Silver-washed Fritillary** – Dog-violet (*Viola riviniana*)

**Small Blue** – Kidney Vetch (*Anthyllis vulneraria*)

**Small Copper** – Mainly Common Sorrel (*Rumex acetosa*) and Sheep's Sorrel (*Rumex acetosella*) and occasionally Dock (*R. obtusifolius*)

**Small Heath** – Fine grasses including Bents (*Agrostis* spp.), Fescues (*Festuca* spp.) and Meadow Grasses (*Poa* spp.)

**Small Pearl-bordered Fritillary** – Dog-violet (*Viola riviniana*) and Marsh Violet (*V. palustris*) and occasionally other violets

**Small Skipper** – Mainly Yorkshire Fog (*Holcus lanatus*). Occasionally: Timothy (*Phleum pratense*),

Creeping Soft Grass (*Holcus mollis*), False Brome (*Brachypodium sylvaticum*), Meadow Foxtail (*Alopecurus pratensis*) and Cock's-foot Grass (*Dactylis glomerata*)

**Small Tortoiseshell** – Stinging Nettle (*Urtica dioica*) and Small Nettle (*U. urens*)

**Small White** – Cultivated brassicas (especially Cabbages and Brussels Sprouts), Nasturtium (*Tropaeolum majus*) and wild crucifers such as Wild Cabbage (*Brassica oleracea*), Garlic Mustard (*Alliaria petiolata*) and Hedge Mustard (*Sisymbrium officinale*)

**Speckled Wood** – False Brome (*Brachypodium sylvaticum*), Cock's-foot Grass (*Dactylis glomerata*), Yorkshire Fog (*Holcus lanatus*), Couch Grass (*Elytrigia repens*)

**Swallowtail** – Milk Parsley (*Peucedanum palustre*)

**Wall** – Tor-grass (*Brachypodium pinnatum*), False Brome (*B. sylvaticum*), Cock's-foot Grass (*Dactylis glomerata*), Bents (*Agrostis* spp.), Wavy Hair Grass (*Deschampsia flexuosa*), Yorkshire Fog (*Holcus lanatus*)

**White Admiral** – Honeysuckle (*Lonicera periclymenum*)

**White-letter Hairstreak** – Elms (*Ulmus* spp.)

**Wood White** – Meadow Vetchling (*Lathyrus pratensis*), Bitter Vetch (*L. linifolius*), Tufted Vetch (*Vicia cracca*), Common Bird's-foot Trefoil (*Lotus corniculatus*), Greater Bird's-foot Trefoil (*L. pedunculatus*)

# Flowers especially loved by butterflies

It is the nectar they drink, and these plants are especially favoured: *Buddleja davidii* varieties (Butterfly Bush), *Lavandula* varieties (Lavender), *Hylotelephium spectabile* (Ice Plant or Sedum), *Knautia arvensis* (Field Scabious), *Verbena bonariensis*, *Centranthus ruber* (Red Valerian), *Hebe* varieties, *Lonicera periclymenum* (Honeysuckle), *Liatris spicata* (Kansas Gayfeather), *Origanum vulgare* (Wild Marjoram), *Centaurea nigra* (Knapweed), *Mentha* spp. (Mint), *Eupatorium cannabinum* (Hemp Agrimony), *Erysimum* (Perennial Wallflower), *Agrostemma githago* (Corncockle), Perennial Salvias, *Monarda* varieties (Bee Balm), *Phlox*, most perennial daisies: *Aster*, *Echinacea*, *Rudbeckia*, *Helenium* and the like.

# Things you can do

— Clip box, privet and other hedges and topiary specimens.
— Make successional sowings on the veg patch.
— Feed plants in containers with liquid feed every week or two.
— Feed border plants and roses once during the month with blood, bone and fishmeal to give them a midsummer boost.
— Remove runners from strawberries.
— Stake tall border plants before they collapse.
— Check the greenhouse for watering morning and evening.
— Remove sideshoots from tomatoes and give the plants a liquid feed once a week.
— Water hanging baskets daily in sunny weather – morning and evening if the weather is really hot.
— Earth up potatoes.
— Plant out tender vegetables such as French and runner beans, courgettes, and outdoor tomatoes.
— Plant out summer bedding in the northern counties and Scotland.
— Feed and train cucumbers and melons in the greenhouse.
— Deadhead roses.
— Water the veg patch in dry weather.
— Pinch the shoot tips from runner beans to deter blackfly.
— Repot auriculas.
— Pot-on hungry house plants into larger containers.
— Divide and replant primroses and polyanthus.

— Cut down any remaining foliage on daffodils and narcissi.
— Harvest early potatoes (they are generally ready when the plants are in flower).
— Prune spring- and early-summer-flowering shrubs, removing a portion of older wood each year.
— Take cuttings of pelargoniums and garden pinks.
— Lightly trim over aubrieta and arabis to prevent them from becoming straggly.
— Dig up, divide and replant overcrowded bearded irises.
— Take time to smell the roses.
— Find the best spot in the garden for a summer evening drink.

# Things you should not do

— Stop cutting asparagus around the middle of the month.
— Don't water the lawn, even in dry spells. It will recover in the first shower of rain and the water is best saved for container-plants and vegetables.
— Don't buy masses of plants unless you know exactly where they are going to go.
— Don't spray to control pests. Let natural predators give a helping hand and grow plants well so they can resist attack.
— Don't go out without sunblock on your nose. With any luck a broad-brimmed hat will be useful, too.

July

The English winter – ending in July,
To recommence in August.

LORD BYRON, *Don Juan*, 1819–24

Sardonic though Lord Byron's opinion might be, there is much to be thankful for in July, for this is the month when the garden becomes an explosion of leaf and flower, provided we have provided our plants with the things they need most – food and water, light and a suitable temperature, though Mother Nature is largely responsible for the latter. The nights are short, the days are long, and on those glorious evenings when the wind has forgotten how to blow and the heady scent of roses hangs on the air, warmed by a day of midsummer sunshine, the view from the garden chair is as beautiful as ever it will be.

High summer we call it and it is as well, on those days when the weather is obliging, to pause, savour the moment and remember why we do this thing called gardening. We do it because it puts us in direct contact with nature; we learn to work with her, to respect her constancy, and when the weather is settled and birds, bees and butterflies go about their business in the hazy days of July, then it seems that all's right with the world. I will add no warning caveat. Summer is here; enjoy it while you can.

# Weather

At last, the only month in the British Isles that is reliably free of frost – even in Scotland. Yes, they've had frosts in June and in August. Hurrah for July! In a cool, temperate climate such as ours, the highs and lows of air pressure can wreak temporary havoc with wind and rain, but the sun is at its height and at last the days are long enough to allow keen gardeners to be up and about before starting their day's work, and to relax with a glass on the terrace come the evening. It was on 19th July in 2022 that the UK recorded its highest ever temperature – 40.3°C – at Coningsby in Lincolnshire. On that day the cool of the evening was especially welcomed and journalists suggested that this would now be the norm as far as British summers were concerned. Time will tell . . .

# Day length (Skipton)

|  | 1 July | 31 July |
| --- | --- | --- |
| **Dawn** | 03.47 | 04.35 |
| **Sunrise** | 04.38 | 05.18 |
| **Sunset** | 21.45 | 21.10 |
| **Dusk** | 22.36 | 21.53 |
| **Daylight hours** | 16 hours 59 minutes | 15 hours 45 minutes |

# Feast days for gardeners

| | |
|---|---|
| **Around 1st Week of July** | **RHS Hampton Court Palace Flower Show** – held in the grounds of the Palace, which means there is plenty of room to walk round and enjoy the floral delights on offer, as well as the setting itself. The great advantage of this show over Chelsea Flower Show is that plants can be bought and carried away from the show ground, rather than just ordered for later delivery. |
| **15 July** | **St Swithun's Day** – St Swithun (also spelt St Swithin) was Bishop of Winchester from 852–862. Should it rain on this day, then folklore tells us that it will continue to rain for forty days and forty nights. The Meteorological Office confirms that since records began in 1861 this has never happened. (Folklore also suggests that if today is sunny, then it will continue fine for the next forty days and nights, but folklorists, being pessimists by nature, often fail to mention that bit.) |

# Tree of the month

## Indian bean tree (*Catalpa bignonioides*)

The true species produces a large, round-headed tree bearing large heart-shaped leaves along with white flowers in summer. Long bean-like pods follow in autumn and may persist on the tree through the winter. You'll need space to accommodate it, but my main reason for choosing it as my tree of the month for July is in its variety 'Aurea', which has leaves of vibrant acid yellow-green that are flushed with purple before they start to expand.

Some might consider it to be a tad strident, but I love it – rather more than trees with dark purple leaves like the copper or purple beech, which can have a lowering effect on the atmosphere. The golden-leafed Indian bean tree is anything but – adding its own brand of sunshine to a garden. Give it space to grow to its full potential and it will be a striking focal point. If you lack the space, then you can cut it back each spring, just before growth starts, and enjoy the fresh young foliage.

Height and spread after 20 years: 15 x 10m.

# Music to listen to . . .

July is the month of roses, so treat yourself to the full-blown emotion of Tchaikovsky's 'Rose Adagio' from *The Sleeping Beauty*. This is a piece that builds in its intensity to the most passionate climax, and if you have ever seen the ballet you will know why. Princess Aurora dances with each of her suitors, who individually offer her a single rose. She accepts them, pirouetting between each, then she stands quite still, *en pointe*, as each suitor, in turn, takes her hand, turns her around, then steps aside leaving her perfectly balanced while the next suitor advances and the audience holds its breath. Having seen Darcey Bussell perform the role, I see her in my mind's eye whenever I hear this glorious music.

If you yearn for calmer fare, there is 'The Flower Duet' from *Lakme* by Leo Delibes, where two soprano voices join in the most glorious vocal confection, or the 'Bailero' ('The Shepherd's Song') from *Songs of the Auvergne* by Joseph Canteloube, both of which seem redolent of those gloriously warm days of high summer.

## Vegetables from the garden this month

Globe artichoke, broad beans, French beans, runner beans, beetroot, calabrese, baby carrots, cauliflower, courgettes, herbs, lettuce, spring onions, onions, peas, early potatoes, radishes, spinach, tomatoes (under glass), turnips.

## Something to muse upon . . .

The year 1945 saw the publication of Flora Thompson's marvellous trilogy *Lark Rise to Candleford*, which was

turned into a television series in 2008. The first part –
*Lark Rise* – originally appeared in 1939 and its stories
tell of Thompson's childhood in the English countryside – Oxfordshire in particular – during the later nineteenth century. They paint a picture of rural life by
turns both tough and picturesque and they demonstrate
how country folk depended almost entirely on the land
for their survival. The books are full of incident and
vivid word pictures; I especially love this extract, of
which I am reminded whenever anyone asks me how
they can best have a labour-saving garden:

On light evenings, after their tea-supper, the men worked
for an hour or two in their gardens or on the allotments.
They were first-class gardeners and it was their pride to
have the earliest and best of the different kinds of vegetables. They were helped in this by good soil and plenty of
manure from their pigsties; but good tilling also played
its part. They considered keeping the soil constantly
stirred about the roots of growing things the secret of
success and used the Dutch hoe a good deal for this
purpose. The process was called 'tickling'. 'Tickle up old
Mother Earth and make her bear!' they would shout to
each other across the plots, or salute a busy neighbour in
passing with: 'Just tickling her up a bit, Jack?'

The energy they brought to their gardening after a
hard day's work in the fields was marvellous. They
grudged no effort and seemed never to tire. Often, on
moonlight nights in spring, the solitary fork of someone
who had not been able to tear himself away would be

heard and the scent of his twitch fire smoke would float in at the windows. It was pleasant, too, in summer twilight, perhaps in hot weather when water was scarce, to hear the *swish* of water on parched earth in a garden – water which had been fetched from the brook a quarter of a mile distant. 'It's no good stintin' th' land', they would say. 'If you wants anything out you've got to put summat in, if 'tis only elbow-grease.' . . .

. . . Most of the men sang or whistled as they dug or hoed. There was a good deal of outdoor singing in those days. Workmen sang at their jobs; men with horses and carts sang on the road; the baker, the miller's man, and the fish hawker sang as they went from door to door; even the doctor and parson on their rounds hummed a tune between their teeth. People were poorer and had not the comforts, amusements or knowledge we have to-day; but they were happier. Which seems to suggest that happiness depends more upon the state of mind – and body perhaps – than upon circumstances and events.

FLORA THOMPSON, *Lark Rise to Candleford*, 1945

# Flower of the month

## Rose

There is no better flower to celebrate the summer than a fragrant rose, whether it be climbing up the house, scrambling through an old fruit tree, pushing up among perennials in a mixed border, or in a bed of its own. But

no, not in a bed on its own. Not any more. I have hoed
the grey earth between hybrid tea and floribunda roses
growing in 'rosebeds' and felt dreadfully sorry for them.
They seem bored and ungainly, with stems shooting out
at odd angles, or else lop-sided and left too long before
pruning.

If you enjoy the bush roses – hybrid teas and floribundas – then prune them well in January or February,
cutting them back to around knee height to maintain
their youth and vigour. Don't snip at them with timidity
so that their stems tower ever upwards and produce
flowers that can only be enjoyed from your first-floor
bedroom windows. All roses deserve company, whether
that is a carpet of hardy geraniums or other ground-
covering plants, or assorted shrubs and perennials in a
mixed border. If you want to admire them on their own,
grow climbers and ramblers on a fence or a house wall,
up an arch or over a pergola and choose varieties that
are fragrant as well as beautiful.

The modern shrub roses (see June) are wonderful in
beds and borders and will flower in fits and starts from
June until October. But I still love the ancient kinds for
their histories and brief but glorious season of flowering
in June and July. The best roses for a wall or fence? I
love 'Golden Showers' for its soft yellow sunshine blossoms, and 'The Generous Gardener' is a warm pink
that frames our bedroom window and flowers massively
in June and July, then intermittently until autumn. Visit
a rose garden this month to see them in their glory and
decide which are the ones for you.

# Flowers in the garden

Abelia, Abutilon, Achillea, Aconitum (monkshood), Agapathus, Ageratum, Allium, Alstroemeria, Althaea (hollyhock), Alyssum (annual), Anaphalis (pearly ever-lasting, Anthemis, Anthericum, Antirrhinum (snap-dragon), Armeria, Artemisia, Aruncus, Aster (annual and perennial), Astilbe, Astrantia (hattie's pincushion), Baptisia, Bedding plants, Bistorta, Borago (borage), Buddleja (Butterfly bush), Butomus (flowering rush), Calceolaria, Campanula (bellflower), Cardiocrinum, Carpenteria, Catalpa (Indian bean tree), Ceanothus (Californian lilac), Centaurea (cornflower; annual and perennial), Centranthus (Valerian), Chrysanthemum (annual), Cimicifuga (bugbane), Clematis, Convolvulus, Cosmos, Cotinus, Crocosmia (montbretia), Cytisus (broom), Daboecia, Delphinium (annual and peren-nial), Dianthus (pinks and sweet Williams), Dicentra, Dictamnus (burning bush), Dierama (Angel's fishing rod), Digitalis (foxglove), Echinacea (coneflower), Echinops (globe thistle), Erica (heather; summer flow-ering), Erigeron (fleabane), Erodium, Eryngium (sea holly), Escallonia, Eucryphia, Euphorbia (spurge), *Fallopia baldschuanica* (Russian vine), Filipendula (meadowsweet, Dropwort), Fremontodendron, Fuchsia, Gaillardia, Galega (goat's rue), Galtonia, Gaura, Genista, Gentiana (gentian), Geranium (cranes-bill), Gladiolus, Gypsophila (baby's breath), Hebe, Helenium (sneezeweed), Helichrysum, Heliopsis,

Hemerocallis (daylily), Heuchera, Hosta, Hydrangea,
Hypericum, Impatiens (busy lizzie), Indigofera, Inula,
Iris, Jasminum (jasmine), Kniphofia (red-hot poker),
Lathyrus (sweet pea), Lavandula (lavender), Lavatera
(mallow; annual and perennial), Leontopodium (edel-
weiss), Leycesteria (himalayan honeysuckle), Liatris
(Kansas gayfeather), Ligularia, Lilium (lily),
Limonium (Sea lavender, Statice), Linaria (toadflax),
Lobelia, Lonicera (honeysuckle), Lupinus (lupin),
Lychnis (catchfly), Lysimachia (loosestrife), Lythrum
(purple loosestrife), Macleaya, *Magnolia grandiflora*,
Malva (mallow), Meconopsis (Himalayan poppy),
Mimulus (monkey flower), Monarda (bergamot, Bee
balm), Morina, Myrtus (myrtle), Nandina (sacred
bamboo), Nepeta (catmint), Nicotiana (tobacco plant),
Nuphar (brandy bottle), Nymphaea (water lily),
Oenothera (evening primrose), Papaver (poppy),
Passiflora (passionflower), Pelargonium, Penstemon,
Petunia, Phlox, Phygelius (cape fuchsia), Physostegia
(obedient plant), Platycodon (balloon flower),
Polemonium (jacob's ladder), Pontederia, Potentilla,
Primula, Rodgersia, Romneya (tree poppy), Rosa
(roses), Rubus, Rudbeckia, Salvia (annual and peren-
nial), Santolina (cotton lavender), Scabiosa (scabious),
Schizophragma, Sedum (ice plant), Sempervivum
(house leek), Senecio, Sidalcea, Sisyrinchium,
Solanum, Solidago (golden rod), Spiraea, Stachys,
Tagetes (French and African Marigold), Tamarix
(tamarisk), Teucrium (germander), Thalictrum,
Thymus (thyme), Tigridia, Trachelospermum (star

jasmine), Tradescantia, Tropaeolum (nasturtium), Verbascum (mullein), Verbena, Veronica, Viola (violet, Pansy), Vinca (periwinkle), Yucca.

# A garden to visit in July

There is one British garden above all others that folk seem to drool over: Sissinghurst, near Cranbrook in Kent. It is not difficult to see why. The surrounding countryside – the Weald of Kent – has its own intrinsic beauty and the garden has been an inspiration since it first entered the public consciousness in the 1950s. Created by Harold Nicolson (a diplomat and diarist who was mainly responsible for the design and layout) and his wife Vita Sackville-West (a passionate gardener and writer), the story of Sissinghurst's evolution is only enhanced by the reputations and predilections of its creators, of which much has been written, not least by their son Nigel Nicolson in his revealing *Portrait of a Marriage*.

Go there in late June or July and marvel at the magic of the place. Beneath the atmospheric Elizabethan tower, in which Vita wrote her gardening books, her columns and her novels, lie a series of gardens and pathways, interlocking, intertwining and bewitching. The famous White Garden is only a small part of an elaborately constructed design, complete with mellow brick walls draped with a tapestry of climbing plants and beds and borders, where you will find yourself drowning in a heady mixture of colour and fragrance.

The garden is now in the care of the National Trust. Head gardeners have come and gone since the days of Vita's chosen pair of Pamela Schwerdt and Sybille Kreutzberger, but right now the place manages to capture the essence of the Nicolson/Sackville-West magic. Don't miss it.

# Sitting down

You know you *want* to do it. You *try* to do it. But it doesn't last. You promise yourself a break from the toil. It is July. The garden is looking at its best. Well, apart from that bit of bindweed that is twining up the stalk of a delphinium and sticking its tongue out at you. Resist the temptation to rise from your seat.

Oh, alright then; get up, untwine it and stick it where it can't do any harm, then sit down and look the other way. There are weeds sneaking through the gaps between the paving. Where is that old dinner knife that is so good at prising them out? It's in the shed. You will go and get it. Now, stop right there and read the following out loud to yourself:

There is absolutely no point in having a beautiful garden if I do not appreciate it and simply admire it from time to time. I must work harder at sitting still, at breathing deeply and inhaling the fragrances of garden pinks and roses. If I do not, all my work might as well have been in vain. Only by looking at the scene I have created will I find a way to true peace of mind. Others call it mindfulness. I call it gardening. The sky is blue; nature (along with blood, bone and fishmeal) has worked its magic. If I do not take time to pause and be thankful for what I have achieved, I am an ungrateful wretch and the angels will weep for me.

ALAN TITCHMARSH, 2022

There. Better? Yes, alright, now go and find that dinner knife . . .

# Bird of the month

## Song thrush (*Turdus philomelus*)

I feel for the song thrush on two accounts: first, its unpleasant Latin name; second, its predilection for flying into windows at its own reflection and ending its life so ingloriously. Its song is heavenly, its plumage an exemplary lesson in tonal beauty – that gloriously spotted breast – and its ability to crack open the shells of snails and eat them is a facility guaranteed to endear it to any gardener. If you do not hear the song thrush singing, you may well hear the 'crack, crack, crack' of that snail shell as it brings it down on the anvil of the paving slab or stone that it uses to get to its tea. As far as its name goes, there are only slightly more euphonius alternatives: Throstle or Mavis. Mmm.

So highly was the bird's song regarded by early settlers in the Antipodes – in the 1860s – that they took thrushes with them. Alas, the bird failed to get a foothold in Australia, but it thrives to this day in New Zealand. The song is amazingly complex, with around 100 variations, and over the years has inspired poets as diverse as Thomas Hardy, Edward Thomas and Ted Hughes. Don't confuse the song thrush (22–24cm) with the larger mistle thrush (26–28cm), seen robbing hedges of berries in the dead of winter. The redwing (20–22cm) is another similar winter visitor with a spotted breast, but the reddish tone beneath its folded wings is a good means of differentiation.

Thrushes feed on worms, insects, snails and berries. They make cup-shaped nests in trees and shrubs and will lay four to six pale blue eggs speckled with black in two or three broods between March and June. The young will fledge in 11–15 days. There are around 1.3 million breeding pairs scattered across the whole of Britain and Ireland.

# Deadheading

'Off with their heads,' cried the Queen of Hearts in Lewis Carroll's *Alice's Adventures in Wonderland*. She might well have been an accomplished gardener, knowing that to remove faded flowers from roses, border perennials and many other plants will save the energy they might well put into forming unwanted seeds, but also – and more importantly in most gardens – it will make beds and borders look neater and less 'over the hill' and may also result in plants producing a second flush of flowers. Only where the seedheads are decorative, or where they will offer sustenance for birds and insects, should they be left to develop. Otherwise, snip them off as soon as they go over the hill and put them on the compost heap.

With hybrid tea and floribunda roses, once all the flowers on a particular stem have faded, cut the stem back into growth that is of pencil thickness, rather than the thin bit just below the flower cluster. The resulting second flush of bloom will appear slightly later in the summer as a result, but it will be more vigorous and composed of more flowers – the thicker stem having a greater ability to regenerate. Plants to avoid deadheading? Those that produce attractive berries, grasses and, obviously, those where the fruits are the main reason for their cultivation.

# Fruit of the month

## Peach

They will be cropping this month, but if you've a warm wall, or even a sunny and sheltered spot on a terrace or patio, it really is worth planting a peach from a container now. The flavour of a sun-warmed fruit, encased in its velvety skin, conjures up memories of holidays in more propitious climes, and with today's new varieties you can even grow a peach in a pot, thanks to dwarfing rootstocks. Traditionally, peaches would have been grown with apricots and nectarines as trained trees against the back wall of a large lean-to greenhouse in the kitchen garden of a stately home, but they will also survive outdoors, planted against a sunny south- or west-facing wall with their branches fanned out and tied in to wires.

'Peregrine' is an old traditional variety with large, delicious fruits, but new dwarf kinds such as 'Avalon Pride' can be grown free standing and will reach a height of around 2m, much less if grown in a large pot or tub. The flowers are self-fertile, but make sure you pollinate the flowers with a soft paintbrush when they open in spring to help with fruit setting, and be ready, outdoors, for peach leaf curl, which causes the leaves to pucker and redden before falling off. Dispose of the infected leaves rather than composting them. The plant will get over the attack and grow through it, particularly if you apply foliar feed once a week from May onwards, which will help strengthen its constitution.

The fungus disease is less of a problem in a greenhouse, but you'll need a large one to grow peaches or apricots (smaller and not quite so juicy), or nectarines (smooth-skinned peaches). When the moment comes to harvest that first fruit – sometimes even in the plant's first year with you – what a thrill!

# Fruit from the garden

Very early apples such as 'Beauty of Bath', apricots, peaches and nectarines from a greenhouse, blackberries, cherries, black, red and white currants, gooseberries, loganberries, plums, raspberries, strawberries.

# Something to read . . .

I am aware that I might begin to sound like the member of the aristocracy who, on addressing Royal Horticultural

Society members some years ago, began his talk with the words: 'No matter how small your garden, always be sure to devote at least four acres to woodland.'

So it is with a degree of self-consciousness that I confess that one of my greatest pleasures is the wild flower meadow I sowed, by hand, some ten years ago and which is now awash with all manner of wild flowers from April through to September. Cowslips turn my meadow (it is 2 acres if you wish to compare me with the woodland advocate) sulphur yellow in spring, then vetches, marguerites, scabious and knapweed follow until the purple haze of marjoram provides a grand finale at the end of summer. In the first week in September, when the seed has fallen, we cut it down, remove the 'hay' and wait for another year.

'Meadows' can be as large or as small as you want. They will attract bees, butterflies and other insects, and give you a break from the more intense style of cultivation that occupies the rest of your garden. November is the best time to sow them, and if you want to read all about them and enjoy inspiring writing at the same time, turn to Christopher Lloyd's *Meadows*, where the illustrations will delight you as much as Christo's delightful prose.

# Famous gardener of the month
## Lancelot 'Capability' Brown (1716–1753)

His nickname came about as a result of saying to would-be clients that their land had 'great capabilities' and he went on to create some of our finest large-scale landscapes, around 150 of which survive. Born and brought up in Kirkharle, Northumberland, and apprenticed as a gardener, Lancelot Brown went on to garden at Stowe and to consort with kings and queens and the nobility and gentry of the day, for whom he designed naturalistic landscapes that brought the countryside right up to the front door of their stately homes.

He used serpentine lakes to give the impression of meandering rivers, and he was the master of 'the borrowed view', the vista and 'the peep' – a glimpse through trees of distant eye-catchers. But his landscapes were practical, too, in that they allowed for the cultivation of woodland and the grazing of livestock, all with the idea of creating an Arcadian idyll based on sound horticultural and agricultural practices. He designed houses as well as landscapes and later in life took charge of the gardens at Hampton Court, where he lived in the Wilderness House and planted the famous grape vine. Chatsworth, Claremont and Stowe show ample evidence of his talents.

The thing that has always struck me about Brown is that he must have been a 'people person' in order to have got on not only with his high-ranking clients, but also the labourers who worked for him year in, year out, on the contouring of land that had to be accomplished with pick and shovel and wheelbarrow before the days of the mechanical digger.

By the time of his death, he had earned, in today's equivalent, around £500 million. Out of that he would have had to pay the workers on around 260 different sites, but the remaining £130 million would still have allowed him and his family an extremely good standard of living – had he sat still for more than a day or two at a time to enjoy it and not spent most of it travelling the country from one estate to another in a horse-drawn carriage; not at all bad for a former apprentice gardener. He died in London as a result of a fall, which it is thought led to a fatal asthma attack. He was sixty-six, and probably absolutely exhausted.

# The compost heap

You will read as many complex narratives about the making of garden compost as you will about what the government should really do about any given political situation. I will have none of it. If life is too short to stuff a mushroom, then it is certainly too short to spend time agonising about the construction and maintenance of a load of old rot. When plant material is removed from its life source – the plant – it's natural inclination is to rot. It is up to the gardener not to get in the way of this natural process. End of story. You do not need to turn the heap every few weeks, or add accelerators or soil or anything else for that matter. But . . . there are a few useful things to remember:

1. Make an enclosure to contain the stuff – posts and wooden sides, posts and wire – anything that stops it slithering around.

2. Put into the heap anything soft and green – vegetable waste, faded flowers, annual weeds, lawn mowings, even woody stems provided that they have been through a shredder (a great buy).

3. Avoid cooked foods – bread, mashed potatoes and the like that will attract rats. Thick-rooted weeds like bindweed are also best omitted in case they survive to infest other parts of the garden.

4. Mix the different kinds of stuff when you put them in so that there is no concentration of any one thing in one spot – especially lawn mowings, which will go slimy.

5. Keep the heap moist by turning a hosepipe on it in dry weather.

6. Tamp it down with a rake to keep it firm as you fill it.

7. Cover it with a piece of old carpet to prevent it from drying out.

8. Have at least two heaps so that one can be filling while the other is rotting (this happens quicker in summer than in winter).

9. Fill a heap during spring and summer and use the rotted material in late winter and spring when you are cultivating soil.

10. That's it.

# Wild flower of the month

## Field scabious (*Knautia arvensis*)

I find the flowers of this plant wonderfully romantic. Perhaps it is the combination of their fragility, the soft blue colouring of their petals and the powdery off-white anthers that pepper the centre of the flower where the cup-shaped petals are paler. It may also have something to do with the fact that my grandfather grew the larger cultivated version on his Yorkshire allotment when I was a boy and cut them for my grandmother, along with savoy cabbages and sweet peas. The field scabious towers over other meadow plants in midsummer at around a metre high and the stems sway elegantly in every passing breeze. On the mown rides in the meadow it still survives to flower at a height of around 10cm between cuts. It is especially happy and prevalent on chalky downland and its flowers attract both bees and butterflies.

*Other local common names*: Bachelor's Buttons, Billy Buttons, Black Soap, Blackamoor's Beauty, Blue Bonnet, Cardies, Clodweed, Coachman's Buttons, Gentleman's Pincushion, Gipsy Rose, Lady's Hatpins, Lady's Pincushion, Mournful Widow, Pins-and-Needles, Robin's Pincushion, Soldier's Buttons, Teddy Buttons.

# Things you can do . . .

— Summer prune trained fruit trees.
— Prune wisteria, shortening unwanted growths to around 30cm.
— Deadhead faded flowers.
— Trim hedges.
— Take cuttings of border plants, pot plants, alpines and the like.
— Liquid feed plants in containers once a week.
— Ventilate the greenhouse (automatic ventilator arms will save you a lot of time and trouble) and check plants for water morning and evening.
— Plant out leeks sown in March.
— Harvest potatoes.
— Make successional sowings of salad crops.
— Feed tomatoes once a week and remove sideshoots from single-stemmed varieties.
— Train cucumbers and feed weekly.
— Dig up and store tulip bulbs – keeping only the large ones.
— Cut down any remaining foliage on daffodils and narcissi.
— Stand citrus fruits outdoors where they will not be scorched by midday sun. Feed them once a week.
— Summer prune cherries and plums.
— Thin out overcrowded apples and pears.
— Clear strawberry beds of straw and cut back the foliage once the fruits have been harvested.
— Make new strawberry plants by pegging down runners

into sunken pots of peat-free multi-purpose compost.
— Prune weigela and philadelphus, removing a portion of older wood.
— Cut and store herbs for winter use by freezing or drying.
— Earth up celery.
— Keep veg crops well supplied with water in dry spells.
— Sow spring cabbages.
— Summer prune trained apple and pear trees.
— Plant colchicums and autumn-flowering crocuses.
— Clip over helianthemums when their flowers fade.

# Things you should not do . . .

— Don't go on holiday without asking a friend or neighbour to keep an eye on watering your containers and harvesting crops that are ready to be picked.
— Don't forget to water newly planted trees, shrubs, bedding and perennials during dry spells.
— Resist watering the lawn. It will soon recover come the rains.
— Don't be a slave to the garden. Enjoy it.

August

August rain: the best of summer gone, and the
new fall not yet born. The odd uneven time.

SYLVIA PLATH, *The Unabridged Journals of Sylvia Plath,* 1982

August is a bittersweet time and, as Sylvia Plath intimates, a month of in-betweenness. High summer arrived with July. September is late summer. August is . . . August, with all of its strange baggage, for the novelty of summer has worn off and autumn is not quite in sight. Perhaps that's why we take holidays this month, so that we can be away when the seasons turn and come back in September knowing that summer has evaporated while we looked the other way.

Or am I being maudlin? I love September, but not for nothing did that great gardener Vita Sackville-West leave her beloved Sissinghurst and go on her travels in August, believing that the best was over. No; we must fight against defeatism this month, and revel in later summer flowers and second flushes of the earlier blooms thanks to assiduous deadheading. Autumn and winter will come soon enough, and with an August harvest of peas and beans, luscious salads, sun-warmed tomatoes and tasty potatoes, we will learn to savour and be grateful for the moment.

# Weather

The likelihood of frost is minimal in August, though in Scotland it is not unknown towards the end of the month with the shortening days and chilly nights. But in general we experience a mixture of weather: hot and sunny, bright and breezy, cloudy and threatening with a reasonable likelihood of thunderstorms. But with luck the outlook will be relatively equable, and those British seaside resorts that depend heavily on tourism will not be disappointed. Let's face it, summer holidays are booked well in advance and Brits have become used to gritting their teeth and facing with relative equanimity those regular holiday delights: sand in the egg sandwiches, queues on the motorway, cries of 'are we there yet', and listening to the sound of rain pounding on the car roof as we gaze at an angry sea.

But then, we can always stay at home and enjoy the garden . . .

# Day length (Skipton)

|         | 1 August | 31 August |
|---------|----------|-----------|
| **Dawn**    | 04.37    | 05.37     |
| **Sunrise** | 05.19    | 06.13     |
| **Sunset**  | 21.08    | 20.03     |

| Dusk | 21.51 | 22.39 |
| --- | --- | --- |
| Daylight hours | 15 hours 42 minutes | 13 hours 44 minutes |

# Feast days for gardeners

| 1 August | **Lammas** – Loaf Mass Day, when the grain harvest is celebrated. A time for making bread and corn dollies and enjoying feasts with friends and family. |
| --- | --- |
| 1 August | **Yorkshire Day** – when all things Tyke are celebrated. Well, I had to put that one in . . . |
| 1st Monday in August | **Bank Holiday** in Scotland. |
| 30 August | **St Fiacre of Breuil** – the patron saint of gardeners celebrates his feast day on 30 August or 1 September. He was an Irish hermit who moved to France where his charitable work and horticultural expertise became renowned. He died in AD 670. |
| Last Monday in August | **Late Summer Bank Holiday** in England, Wales and Northern Ireland. |

# Holidays

How we look forward to them! After all, half the fun of having a holiday is in the anticipation of its arrival. (The other half is watching other people still working when we are not – so opined Kenneth Grahame in *The Wind in the Willows*.) Until 1834 there were thirty-three public holidays in the UK, but in that year the number was deemed excessive and greatly reduced. We now have only eight public holidays a year: New Year, Good Friday, Easter Monday, May Day, late May bank holiday, August bank holiday, Christmas Day and Boxing Day.

Why 'bank holidays'? That's thanks to the Liberal politician and banker Sir John Lubbock, who authorised the Bank Holidays Act of 1871, making the four days official. For a few years the four days were referred to as 'St Lubbock's Day'. (For some unaccountable reason the sobriquet did not stick.) The late May bank holiday replaces the traditional holiday of 'Whit Monday' – the day after 'Whit Sunday' or Pentecost, which falls seven Sundays after Easter Day. The change was made in 1971 so that the holiday always falls on the last Monday in May, but the holiday was never given a name. As a result we simply call it the Spring bank holiday.

In 2022 we enjoyed extra bank holidays in early June to celebrate the Platinum Jubilee of Her Majesty Queen Elizabeth II.

# Tree of the month

## Eucryphia (Tasmanian leatherwood)

Not as widely grown as it should be, eucryphia is a handsome evergreen or deciduous shrub or small tree with dark, glossy green pinnate leaves and – in late summer and autumn – masses of white flowers with shaving-brush centres. Generally regarded as a touch on the tender side, I have seen the deciduous *Eucryphia glutinosa* growing in Beatrix Potter's garden at Sawrey in Cumbria and it is pretty chilly there in winter. *E.* x *nymansensis* 'Nymansay' is even more spectacular with its evergreen leaves and slightly larger flowers of wonderful delicacy. I am also enamuored of Eucryphia lucida (pink cloud); an evergreen variety with glorious blushing flowers.

Give eucryphia a sheltered spot where it is protected from cold winds. Rather like clematis, it prefers its head in the sun and its roots in the shade where they can enjoy the coolness. The soil needs to be moisture retentive and acid rather than alkaline. The plant will look sickly grown in chalky soil, but you can always plant it in a large container of lime-free ericaceous compost and enjoy it as a shrub rather than a tree. I've always been fond of this plant, not least because it comes into flower when many other trees and shrubs have faded from glory. Give it a try and see if you don't agree.

Height 12m; spread 4m (only after 20 years or more).

# Music to listen to . . .

With the summer in its deepest moment, now is a good time to revel in Beethoven's *Symphony No.6* – the Pastoral – with its glorious lush imagery of the British countryside. If foreign holidays are out of the question, and musical warmth will be much appreciated, Rodrigo's Guitar Concerto or Chabrier's *España* will lift the spirits. So will the *Florida Suite*, a glorious confection by Frederick Delius, who might have been born in Bradford but who found the warmth of the Sunshine State more to his liking than the more temperate climes of the West Riding of Yorkshire.

Lighter fare comes in Cole Porter's score for *Anything Goes*, where you can climb aboard the cruise ship and escape with some of his best music. Then there is George Gershwin's luscious *Rhapsody in Blue* – as lush and lyrical as August seems to demand.

# Vegetables from the garden this month

Globe artichoke, broad beans, French beans, runner beans, beetroot, broccoli (calabrese), cabbage, carrots, cauliflower, celery, chilli peppers, courgettes, cucumbers, endive, kohl rabi, lettuce, marrows, mustard and cress, onions, peas, sweet peppers, potatoes, radish, spinach, sweet corn, tomatoes, turnips.

# Something to muse upon . . .

The artist and photographer Sir Cecil Beaton lived for many years at Reddish House in Wiltshire. In addition to his photographic and artistic skills, he was an accomplished diarist – witty, observant, and as creative and inspiring with words as he was with stage sets and costume designs. He loved his garden, even if he did bemoan the fact that he had to be away from it for long periods. His evocation of high summer is wonderfully atmospheric:

This is the day that compensates for weeks of winter cold and rain. The sky is unfettered azure. The trees are still salad-green, and the lawns and fields like Ireland (as well they should be after the months of rain we have had); the birds twittering, piping, chirping, chirruping and hiding. The hoots of the owl, which generally augur more rain, cast no jarring note, for the barometer shows that it will remain fine – if not forever, at least for tomorrow.

On the terrace butterflies hover over the heliotrope, nepeta, and tobacco flowers. Flapping back and forth their wings, they are relishing the pollen with greed and unction. They congregate particularly on the flat plates of the sedum or ice plant – red admirals, cabbage whites and chartreuse yellows. They take their fill and then fly off, possibly through the muslin curtains of my bathroom where they flutter frantically from window to window, beating against the panes until my huge hand clutches at their powdery, panicky wings and imprisons them. Then, too terrified to move in the dark palm of my closed fist, they suddenly find themselves unbelievably fantastically free to fly to the top of the walnut tree or, even further, to the row of elms beyond.

It is a poignant, nostalgic time of year with the knowledge that it cannot last forever. As a flame flickers brightest before it finally goes out, summer is putting on its finest spurt. But it cannot be long before the frost will gradually scythe everything before it, and the garden will go into hibernation for all those long partridge-coloured months. Already the early morning produces a thick, silver-skeined carpet of cobwebs – for the dew is heavy; even at midday it has only been partially consumed by the sun.

CECIL BEATON, *The Restless Years: 1955–63*

# Flower of the month

## Dahlia

Despised as being 'common' and 'vulgar' by many plant snobs during the latter half of the twentieth century, dahlias have, mercifully, been welcomed back into the fold and it is easy to see why. Yes, they are brash and blowsy, but they lift the spirits, brighten the garden and are a great way of prolonging its season of interest, since they flower non-stop from July right through until the frosts of October or November. This ability to 'up the ante' in late summer is one of their greatest attributes, along with the fact that they are great for cutting. If you want massive blooms for the local flower show, grow the 'large decorative' or 'cactus' types – the latter with quilled petals. If you remove the side buds on each flower stalk as soon as they are large enough to handle, leaving just the central one, you will produce larger blooms.

Dahlias (named after the Swedish botanist Anders Dahl) grow from tubers that are planted in the garden in April or early May. Alternatively, rooted cuttings can be planted out in late May when danger of frost is past. Most of them will need staking and a row or two on the allotment or veg patch will mean you can cut the blooms without feeling guilty. They insist on full sun and love rich soil and plenty of moisture. The colour range encompasses red, orange, yellow, purple, white, pink . . . almost everything except a true blue. Dig up the tubers once the leaves and shoots have been blackened by frost and store them in a cool, dark, frost-free shed or garage until the following spring. Alternatively, mulch them heavily and risk leaving them in the ground. In milder areas they will often survive to emerge again the following year.

# Flowers in the garden

Abelia, Acanthus (bear's breech), Achillea, Aconitum
(monkshood), Allium, Alstroemeria, Althaea (holly-
hock), *Amaryllis belladonna*, Anaphalis (pearly everlast-
ing), *Anemone hupehensis* and *Anemone* x *hybrida*
(Japanese anemones), Anthemis, Antirrhinum (snap-
dragon), Aponogeton, Artemisia, Aster (annual and
perennial), Astilbe, Bedding plants, Berberidopsis,
Bistorta, Buddleja (butterfly bush), Campanula,
Campsis (trumpet vine), Canna (Indian shot),
Caryopteris, Catananche (Cupid's dart), Ceanothus
(Californian lilac), Centaurea, Ceratostigma,
Chrysanthemum (annual and perennial), Cimicifuga
(bugbane), Clematis, Clerodendrum, Colutea,
Convolvulus, Coreopsis (tickseed), Corydalis, Crinum,
Crocosmia (montbretia), *Cyclamen hederifolium*, Cytisus
(Broom), Desfontainea, Dianthus, Dierama (angel's
fishing rod), Echinacea, Echinops, Erigeron, Erodium,
Escallonia, Eucryphia, *Fallopia baldschuanica* (Russian
vine), Filipendula, Fuchsia, Gaillardia, Galega (goat's
rue), Gaura, Genista, Gentiana (gentians), Geranium,
Geum, Gladioli, Gypsophila (baby's breath),
Hedychium (ginger lily), Hedysarum, Helenium,
Helianthus, Hemerocallis (day lily), Heuchera,
Hibiscus, Hosta, Hydrangea, Hypericum, Indigofera,
Inula, *Jasminum officinale*, Kniphofia, Koelreuteria
(golden rain tree), Lathyrus (Sweet pea – annual and
perennial), Lavatera (mallow, annual and perennial),

Leycesteria (Himalayan honeysuckle), Liatris (Kansas
gayfeather), Ligularia, Lilium (lilies), Limonium,
Linaria (toadflax), Linum (flax), Lithospermum,
Lobelia (annual and perennial), Lonicera (honeysuckle),
Lychnis (catchfly), Lysimachia (loosestrife), Lythrum
(purple loosestrife), Malva (mallow), Mimulus (monkey
flower), Monarda (Bergamot, Bee balm), Nepeta
(catmint), Nymphaea (water lily), Oenothera (evening
primrose), Papaver (Iceland poppy), Penstemon,
Perovskia, Phlox, Phuopsis, Phygelius (cape fuchsia),
Platycodon (balloon flower), Potentilla (herbaceous and
shrubby), Rhodohypoxis, Romneya (tree poppy), Rosa
(roses), Rubus, Rudbeckia, Salvia (annual and perennial), Sanguisorba, Saponaria (soapwort), Scabiosa
(scabious), Sedum (ice plant), Sidalcea, Silene (catchfly), *Solanum crispum*, Solidago (golden rod), Sophora,
Sorbus, Stachys, Stewartia, Thalictrum, Tigridia,
Trachelospermum (star jasmine), Tradescantia,
Tropaeolum (annual Nasturtium and perennials),
Veratrum, Verbascum (mullein), Verbena, Veronica,
Viola, Zauschneria.

# Smuggling

There was a tradition, years ago, espoused by many
keen gardeners while on their overseas travels, that
snippets of this and that could be snaffled from foreign
gardens and brought home – often in a sponge bag that
provided ideal moist conditions – and the cuttings

rooted in the cosy confines of the propagator in the greenhouse at home. It did no one any harm, did it? If cuttings were taken carefully, so that the parent plant was not disfigured, who was to know? We all did it. But we have since learned of the dangers on introducing foreign pests and diseases to this country and 'plant smuggling' is a prime cause.

It may be a slender strip of water separating us from mainland Europe, but it has stood us in good stead over the years, acting like a moat and preventing the arrival of many a foreign bug and blight. At the time of writing, the government's chief plant health scientist tells me that there are now around 1000 deadly pests and diseases threatening to cross the channel and infect – and probably devastate – our plants. This is not some alarmist propaganda, but a real and present danger lurking but a few miles from our shores. One pathogen – xylella – is a deadly bacterial disease, spread by sapsucking insects. It kills olives, lavender, rosemary and a myriad other plants. We have kept it out . . . so far.

Do not take cuttings and pop them in your sponge bag. Take photographs, and buy a packet of seeds on your return. Do your bit to keep our gardens safe.

# A garden to visit in August

The National Garden Scheme 'Yellow Book' will offer rich pickings in August, and touring holidays – on bike or in car – will take you to unfamiliar areas and introduce you to new gardens. For me, the best gardens – whether large or minuscule – are those that reflect the personality of their owner, and Great Dixter at Northiam in East Sussex is one such. Few gardening writers had the charisma of Christopher Lloyd, and the garden created by him – based on a garden begun by his mother and father – still reflects his enthusiasm, his daring and his plantsmanship.

Great Dixter is managed by Fergus Garrett, who was Christo's head gardener for a number of years. This has ensured that the spirit of the owner lives on, but the garden continues to develop and inspire, rather than remaining as some sterile monument to its former glory. Enjoy the formal areas that are informally planted with colour combinations that are often eye-popping, the meadow garden rich in wild flowers, including orchids, and the beds and borders that are sometimes host to as many as three different schemes within twelve months.

A visit to Dixter is enriching at any time of year, but August is especially good for dahlias and other vibrant denizens of one of Britain's most exhilarating gardens. You can buy plants there, too, and take your own bit of Dixter home.

# Watering

Of all the August jobs that give heartache, watering is perhaps top of the list – not least when you are away.

Greenhouse plants, containers and hanging baskets dry out with ludicrous rapidity in hot sunshine, which means that they need to be checked for water morning and evening. Plants that are short of water become stressed; their growth slows down or stops and they wilt. A good soak will usually revive them (especially if it is administered in the evening, allowing them to recover in darkness when there is little demand on their resources), but it is best to avoid them getting to that stage.

With pot plants, a glance at the surface of the compost will usually reveal its moisture content, as will lifting the pot and testing its weight. (You will become accustomed very quickly to the relative heaviness of a dry or a damp potful.) Soak the compost thoroughly and water it again just as it begins to dry out. Apply liquid feed only to damp compost so that it can go straight into action. Greenhouse shading is essential in summer to slow down the drying-out process and to prevent brilliant sunshine scorching flowers and leaves. Blinds that can be raised and lowered are the best option, but if the greenhouse lacks this sophistication, proprietary shading compounds can be painted on to the glass in May and removed in September to increase all-important autumn and winter light levels.

In the garden, confine your watering to plants in containers, anything that has been recently planted, plus fruit and vegetables. Do not worry about established plants unless they wilt, and forget about watering the lawn. It is a waste of a valuable resource, and come the next shower it will soon green up. Fit water butts to the bases of downpipes to catch rainwater (which always seems to be appreciated far more than the stuff from the tap) and when you have to water a bed or border with a sprinkler, leave it on in any one spot for a good half hour.

Old wives' tales suggest that you should not water in bright sunshine. If the plant is in distress and wilting badly, you will do more harm by ignoring its cry for help than you will by giving it a good soak. And remember, it is the roots that need the water, not the leaves and the flowers. Get the water to the point where it can be most useful. And if you are to be away this month, for goodness sake ask a friend or neighbour to help out with the watering in return for a bumper harvest from your veg patch. (No veg patch? A decent present, then, on your return. Dinner in a Michelin-starred restaurant usually goes down well. Something better than 'A Present from Blackpool' anyway.)

# Bird of the month

## Chaffinch (*Fringilla coelebs*)

The chaffinch is always welcome at the bird table thanks to its cheery song – 'pink!' – and the smoked salmon plumage of the male. The female is more muted in shades of buff and brown, but both have wings striped with white. It's easy to understand why the phrase 'as gay as a chaffinch' came about thanks to the bird's plumage, its song and its general demeanour. A friend of mine likens the longer song of the bird (repeated as many as 3000 times in a day) as a bowler's run up to the wicket – a simile that has stayed with me. In Victorian times the birds were caged and prized for their singing. There were competitions in taverns and the best birds could change hands for enormous sums of money.

The chaffinch occurs in all parts of the British Isles and makes its cup-shaped nest low down in shrubs and vegetation. There are one or two broods in April and May consisting of four or five pale blue eggs scribbled with red. The eggs hatch in just under two weeks and the youngsters fledge a fortnight later. Food consists of seeds and fruits and small invertebrates foraged from trees – especially oaks – when feeding its young.

The British population amounts to some 7.5 million breeding pairs.

# Fruit of the month

## Raspberry

The only disadvantage of the raspberry, as far as I am concerned, is that the seeds get stuck between your teeth. But then it's a small price to pay for such delicious, succulent fruits. There are two distinct kinds of raspberry – those that fruit on stems (called canes) made the previous season (the so-called summer-fruiters) and those that fruit on canes that have grown in the current spring and summer and which carry their fruits rather later as a result – often known as autumn fruiters, though the crop ripens much earlier than that. I grow both types, since my grandchildren will pick raspberries far more willingly than strawberries, as they do not have to kneel down and go ferreting about among the leaves to find them. Try 'Malling Jewel' if you want a tasty summer fruiter; 'Autumn Bliss' for a later crop.

The advantage that raspberries have over strawberries is that they freeze well, so there is no need to be embarrassed by a glut. Plant new canes in autumn and be ready to replant plantations after a few years, since they will eventually succumb to yield-reducing virus diseases. Summer fruiters are pruned immediately after cropping, cutting out all canes that have borne fruit and tying in the new ones. All the canes of autumn fruiters are cut down to the ground in late winter. You'll find wild rasps growing in woodland, which is a useful indicator that they don't mind growing in dappled shade.

# Fruit from the garden

Early apples such as 'George Cave', apricots, peaches and nectarines, blackberries, cherries, black, red and white currants, figs, gooseberries, grapes, loganberries, melons, mulberry, early pears such as 'Doyenne d'Ete', plums, raspberries.

# Something to read . . .

You'll have gathered by now that my reading tastes are
pretty catholic, but all the books I am recommending in
the almanac are anchored in the great outdoors and
should strike a chord with anyone who has an affinity
with nature and the countryside. I write novels myself
(try *The Gift* – the story of a Yorkshire shepherd boy
with an ability to heal animals and people), but I read
other folks' tales with avidity. I was especially entranced
by *Where The Crawdads Sing* by Delia Owens – the
evocatively titled story of a girl who grows up alone, for
the most part, on the marshes of North Carolina. It is a
novel with the ability to draw the reader in totally to a
part of the world where nature has a stronghold. Read it
and be transported . . .

# Famous gardener of the month

## Kim Wilkie (1955–)

Of all our current landscape architects, Kim Wilkie
stands out as being one of the most innovative. When it
comes to recognising what is often referred to as 'the
genius of the place' he has few equals, producing
designs that are sympathetic as well as refreshing.
Wilkie's ability to translate his thoughts into the
contours of the earth produces results that are always
striking, but at their heart is a love of nature and the

countryside and a passion to avoid squandering our environmental inheritance.

His feelings for earth and water are at the very heart of his being; his creations are often surprising land-forms (just look at Orpheus at Boughton House in Northamptonshire, for example), but they also possess a sensitivity to the existing landscape even though they occasionally seem to contrast with their surroundings. Wilkie's book *Led by the Land* distills his thoughts, his pre-occupations and his passion for sculpting the land far better than I can do here. What I so admire is his passion for the earth and his unfailing eye for form and detail. Oh, and the fact that he runs a farm as well.

# Sitting down . . . again . . .

Yes, I know, I have moaned already that this is some-thing gardeners do not do nearly often enough. Stop regarding it as laziness and take it for what it is – a pause in your day and a chance to assimilate your garden's beauty and – yes – its shortcomings. Only by sitting down and looking at it for long enough will you realise how it can be improved. You will also very quickly discover that some seats, while beautiful to look at, are not exactly the most comfortable things on which to relax or even perch.

Before you buy a garden bench or chair, test it! Sit in it. Lie on it. Do all those things you would when buying a new bed. It is all very well having a piece of furniture that looks good as a focal point at the end of a path or vista, but if it leaves you with an aching back and a pain in the bum, you might just as well buy a statue. Seats and benches do exist that are comfortable as well as handsome. Choose one that is a combination of the two and you might actually use it. That way, you will benefit from its presence every bit as much as your garden. (And you'll be glad I nagged you.)

# Wild flower of the month

## Wild marjoram (*Origanum vulgare*)

Although this plant might seem to be more at home in the kitchen garden than the meadow, it is one of the late summer delights of chalky grassland – just the conditions that prevail in my own wild flower meadow. When the spring and early summer flowers are over, the meadow takes on a purple haze, thanks to the marjoram. It is seldom as fragrant as it is on the Continent since our temperatures here are cooler, but it is still useful in the kitchen and its clusters of flowers atop 30-cm high domes of leaves brighten the meadow until early autumn. They are usually a dusky purple-pink, but there are occasional paler forms and a few that are white. I enjoy the fact that it blooms late in the season, when a perennial meadow would otherwise be a raggle-taggle mixture of greens.

*Other local common names:* Organ, Organy, Joy of the Mountain.

# August aphorisms . . .

There are a couple of quotations about August that seem to me well worth sharing:

> These late August mornings smelt of autumn from day-break till the hour when the sun-baked earth allowed the cool sea breezes to drive back the then less heavy aroma of threshed wheat, open furrows and reeking manure.

COLETTE, *The Ripening Seed,* 1923

> No one can believe that God is not good when the August gardens are in their heyday.

GLADYS TABER, *The Book of Stillmeadow,* 1948

# Things you can do . . .

— Order spring-flowering bulbs.
— Check containers for watering morning and evening.
— Take cuttings of pelargoniums and root them in sandy compost.
— Continue to deadhead roses and border flowers.
— Trim fast-growing hedges such as privet and *Lonicera nitida*.
— Take semi-ripe cuttings of shrubs and root them in a propagator.
— Ventilate the greenhouse well and damp down the floors each morning.
— Blanch leeks by earthing up soil around the stems.
— Bend over the foliage of onions to encourage ripening.
— Make successional sowings of salads such as lettuce, radish and spring onions.
— Feed tomatoes, cucumbers and peppers every week.
— Remove sideshoots from tomatoes and 'stop' the plants when they have four or five trusses of fruits.
— Summer prune trained apple and pear trees by pinching out sideshoots.
— Pot up freesia corms for winter colour.
— Sow cyclamen seeds.
— Sow spring cabbages.
— Prune summer-fruiting raspberries by cutting out the fruited canes.
— Start nerines into growth by watering them a little.

— Make new strawberry beds.
— Pot up early narcissi such as 'Soleil d'Or' and 'Paper White', but keep them cool and dark.

# Things you should not do . . .

— Do not put off going on holiday. Ask a neighbour to take over the watering. You need a break.
— Avoid filling the garden with too many small pots, which will be a pain to water.
— Don't water the lawn, however dry it looks. It is a waste of water. Yes, I know I'm repeating myself, but it's important.
— In dry spells, avoid cutting the lawn too short, which will make it even more susceptible to drought. Leave the grass a little longer and it will remain greener.

September

O sweet September, thy first breezes bring
The dry leaf's rustle and the squirrel's laughter,
The cool fresh air whence health and vigor spring
And promise of exceeding joy hereafter.

GEORGE ARNOLD, *September* 1871

September is such a poignant month; not really summer and not really autumn – a sort of in-between period, which comes enveloped in wistfulness for times past and anticipation of what's to come. September offers the same feelings of transition as May, but for quite different reasons. 'Where has the year gone?' we ask ourselves. Since April, the year seems to have been screaming downhill without brakes. And we know we have felt like this before. Every year at this point we look over our shoulder at the months gone by and muse not only upon where they went, but whether the garden has been better than before or not quite up to scratch.

We must think positively. There will, after all, be another spring in only six month's time. All is not lost; if we have planted for our late summer pleasure, there is much to admire still in the late, late summer garden.

# Weather

The occasional baking hot days of July and August are seldom replicated in September. But the weather at this time of year is often more settled and the days not too short, even if the nights are certainly drawing in a little. The schools go back this month (earlier in Scotland) and the cynic would say that's why the weather has picked up – the summer holidays being officially over. North of the border the midge count will start to decline – being at its height between May and early September; prices of bed and breakfast will be reduced right across our islands, and older gardeners, with no need to take their holidays between school terms, will feel suitably smug that they can get away when the weather is equable and the numbers clamouring for accommodation have declined. Ah, the benefits of mature years.

# Day length (Skipton)

|  | 1 September | 30 September |
| --- | --- | --- |
| **Dawn** | 05.39 | 06.33 |
| **Sunrise** | 06.14 | 07.06 |
| **Sunset** | 20.01 | 18.49 |
| **Dusk** | 20.36 | 19.22 |
| **Daylight hours** | 13 hours 40 minutes | 11 hours 37 minutes |

# Feast days for gardeners

| | |
|---|---|
| **1 September** | the start of **Meteorological Autumn.** |
| **1 September** | **Feast day of St Fiacre** – patron saint of gardeners – if not celebrated on 31 August (see August for details). |
| **23 September** | **The Autumn Equinox.** |
| **29 September** | **Michaelmas Day** is the **Feast of St Michael the Archangel** – the triumph of good over evil (St Michael slew the devil in the form of a dragon) – and a reminder that Michaelmas daisies will now be in bloom: 'The Michaelmas daisies among dead weeds, Bloom for St Michael's valourous deeds.' Eat your blackberries before Michaelmas Day, since after that date they are said to have been spat upon by the devil and will taste foul. The Cornish believe that the devil urinated on them, but there is no evidence to prove that late harvested Cornish blackberries or brambles are any more unpleasant than those in other counties. |

# Tree of the month

## *Magnolia grandiflora*

The common name of bull bay does this tree no service at all. It is a glorious evergreen, the undersides of whose shiny, lacquered leaves may be heavily or lightly coated in rufous felt. The flowers – great white waxy goblets that open fully and are scented of lemon mousse – are wonderfully spectacular and are produced throughout the summer and on into late September.

This magnolia loves growing against a warm and sunny wall where some of its branches can be tied in, or out in the open in areas that are not too cold or exposed. It prefers acid soil to those on chalk and has several different varieties, 'Exmouth' being one of the most popular. Unusually for a magnolia it does not resent pruning, neither does it look ugly when it has been cut back sensitively – preferably in spring – to keep it within bounds. Some varieties grow less vigorously and are best for smaller gardens. Look out for 'Little Gem' where space is limited, since it is much less towering of stature.

Height and spread: 15 x 8m (spread less if pruned).

# Music to listen to . . .

You must forgive my apparent self-promotion here. A few years ago I worked with the composer Debbie Wiseman to produce an album of music and poetry called *The Glorious Garden*. I wrote and read poems about different plants and flowers and Debbie composed a piece of music for each. Our album shot straight to No. 1 in the classical music charts due, quite obviously, to Debbie's delightful music. To add to our delight and astonishment, in 2022 the album came in at No. 4 in Classic FM's top 300, the 'Hall of Fame', behind Vaughan Williams' *The Lark Ascending*, his *Variations on a Theme of Thomas Tallis* and Rachmaninoff's *Piano Concerto No.2*.

Debbie's music is worth listening to with or without my verses. Especially popular are: *Myrtle*, *The Snowdrop*, and *The Waterlily*. Listen to them and you will see why the album became so popular.

# Vegetables from the garden this month

Globe artichokes, French beans, runner beans, beetroot, broccoli, cabbages, carrots, cauliflower, celeriac, courgettes, cucumbers, endive, kohl rabi, leeks, lettuces, marrows, parsnips, peas, potatoes, radishes, spinach, swedes, sweet corn, tomatoes, turnips.

# Vegetables in store

Beetroot, onions, shallots.

# Something to muse upon . . .

As I get older I find increasing fascination in diaries; not my own, which shamefully amount to no more than lists of things that need to be done and appointments that need to be kept, but those of interesting people who have lived in interesting times (aren't they always, if you know where to look and when to listen).

At the top of my list of 'best diarists' is James Lees-Milne, Country Houses Secretary of the National Trust, who worked for them from the 1930s to the 1970s and whose marriage to the garden writer Alvilde Lees-Milne was after the fashion of Nigel Nicolson (another good diarist) and Vita Sackville-West. Lees-Milne's diaries – all twelve volumes of them covering the years from 1942 to 1997 – are not simply about his work at the Trust, but are observations of people met and places visited. They are, by turns, amusing, wasp-ish, petulant, irritating, sentimental and perceptive, but always hugely readable.

If the witterings of the upper classes drive you nuts, leave them be. (I admit to sometimes wanting to throw a volume across the room – but I invariably pick it up again and dive in.) If you find the milieu of Kings and Queens, Dukes, Duchesses, and 'the ton' fascinating, work your way through the volumes. The great thing about diaries is that they come in bite-sized chunks, perfect for bedtime reading. My two selections relate to September and delight rather than offend:

Season of mists and mellow fruitfulness. Beautiful Indian summer. The sun takes hours to force its way through, but it is then very hot, and the dogs pant when I take them blackberrying in Westonbirt. Poor little Honey is in decline, suffering from heart murmur. The other darling unaffected as yet, in fine fettle. Last night A. and I leant out of my bedroom window and heard one owl talking to another, somewhere in the garden of the big house. The moon was full, percolating through the cedar trees. We thought it moving to hear an owl hoot – to such a pass have things come, with the elimination of natural creatures.

JAMES LEES-MILNE, *Beneath A Waning Moon*, September 1985

And glorious harvest moon, too, like a silver-gilt porringer by Paul Lamerie. I never remember a more beautiful summer than this has been. September and October are always the best months.

JAMES LEES-MILNE, *Ancient as the Hills*, September 1973

# Flower of the month

## Hydrangea

There are those who think the hydrangea is common (I can't think that James Lees-Milne would have liked them), but I relish their luscious, succulent foliage and big, fat, late summer flowerheads, especially in gardens by the sea where they seem to thrive in the crystal clear air.

It is the mopheads that are most often seen in shades of white, pink and blue (the latter only on acid, lime-free soil). Greater delicacy is possessed by the earlier flowering *Hydrangea arborescens* 'Annabelle'. Her blooms are lime green at first, turning to white, then pale green again, and she does need some kind of corset to hold her up. If you want more eye-popping spectacle, then do seek out *H. paniculata* 'Vanille Fraise', which is totally self-supporting and has conical flower heads that are pale cream flushed with pink – each one like some horticultural ice-cream cornet decorated with sarsaparilla.

Deadhead the mophead hydrangeas in late winter when you can see the fat flower buds forming immediately behind the faded flowers. Take great care to leave these buds intact. Prune the Paniculata hybrids in March or April, cutting them back by about a half – unlike the mopheads they flower on new wood. So does 'Annabelle', whose stems can be thinned out in number and, again, cut back by half their length. Hydrangeas love water (the clue is in the first half of their name) so grow them in rich, moisture-retentive earth.

# Flowers in the garden

Abelia, Achillea, Aconitum, Agapanthus (African lily), *Amaryllis belladonna* (Jersey lily), Japanese anemones,

Asters (annual and perennial Michaelmas daisies),
Bedding plants, Buddleja (butterfly bush), Bupleurum,
Campsis (trumpet vine), Canna (Indian shot),
Caryopteris, Catananche (cupid's dart), Ceanothus
(Californian lilac), Ceratostigma, Chrysanthemums,
Cimicifuga, Clematis, Colchicum (autumn crocus),
Coreopsis (tickseed), Cosmos, Crocus (autumn-flowering species), Daboecia, Echinacea, Erica (heather),
*Fallopia baldschuanica* (Russian vine), Fuchsia, Galega
(goat's rue), Gentiana (gentians), Geranium, Geum,
Gypsophila, Hebe, Helenium (sneezeweed), Helianthus
(sunflower – annuals and perennials), Hibiscus, Hosta,
Hydrangea, Hypericum, Indigofera, *Jasminum officinale*,
Kirengeshoma, Kniphofia (red-hot poker), Lavatera
(annuals and perennials), *Leucojum autumnale*,
Leycesteria (himalayan honeysuckle), Lilium (lily),
Liriope (lily turf), Lobelia (annuals and perennials),
Lonicera (honeysuckle), Lysimachia (loosestrife),
*Magnolia grandiflora*, Monarda (Bee balm), Nerine
(Guernsey lily), Oenothera (evening primrose),
Osmanthus, Papaver (poppy), Pelargonium, Penstemon,
Phlox (alpine and border), Phygelius (cape figwort),
Potentilla (perennials and shrubby types), Romneya
(California tree poppy), Roses, Rudbeckia, Salvia
(annual and shrubby), Scabiosa (scabious), Schizostylis
(river lily), Sedum (ice plant), Solidago (golden rod),
*Sophora japonica*, Tamarix (tamarisk), Teucrium
(germander), Tradescantia, *Tropaeolum speciosum*, Ulex
(gorse), Verbena, Veronica, Vinca (periwinkle), Viola,
Yucca.

# Ornamental fruiting trees and shrubs

Berberis (barberry), Callicarpa (beauty berry), Celastrus, Cotoneaster, Crataegus (hawthorn), Euonymus (spindle tree), Hippophae (sea buckthorn), Malus (crab apples), Pernettya, Pyracantha (firethorn), Roses (hips), Sambucus (elder), Skimmia, Sorbus (mountain ash, Whitebeam), Viburnum, Vitis (vines).

# Garden of the month
## Tresco Abbey Garden, Isles of Scilly

Have you been to paradise? If you've ever visited the island of Tresco, you'll know what I mean. It is a gardener's dream – a small island set in a turquoise sea with a garden planted up with all manner of exotic things, many of which are too tender to be grown on mainland Britain. There is seldom a frost in Tresco. The Dorrien-Smiths have been resident since the island was leased to their ancestor by the Duchy of Cornwall in the mid-nineteenth century.

You can stay here if you wish, in an apartment or cottage, or else take a day trip across the water from Penzance on the *Scillonian III* (2 hours 45 minutes to St Mary's and then a short trip on a smaller boat) or a fifteen minute flight by helicopter. If the former, pray for a calm crossing (not for nothing do the locals refer to the vessel as 'The Great White Stomach Pump'), but on a smooth sea you will enjoy the cruise and wonder what all the fuss was about. Either way this is one garden not to be missed, and in September, after a summer of growth, the garden will be in fine form and rather quieter than in the holiday months.

# Spring-flowering bulbs

Gardeners are always thinking ahead, but pleasurably rather than anxiously. September is spring-bulb planting time. The sooner you get daffs and narcissi in the ground (or in pots) the better. If you've ever speared one with your garden fork during summer cultivations – and brought it to the surface rather like a pickled onion out of a jar – you'll have noticed that new roots have already begun to form. The sooner they are planted (to three times the depth of the bulbs itself) the better. They look best when planted in clumps of ten or twelve, allowing about 5cm between the individual bulbs. Tulips can wait until October or November. I love planting them in containers – eight or ten bulbs to a 30-cm diameter pot.

Try a few new things each year. I cherish camassias with their spires of clear blue starry flowers. They may only last two or three weeks in April and May, but they are great when planted in a meadow where they look most at home. Tulips fizzle out after a year or two when naturalised (planted in grass), but provided they are planted deeply enough and not allowed to dry out in spring, daffodils and narcissi are more reliable.

# Bird of the month

## House sparrow (*Passer domesticus*)

Canaries are brighter of plumage and more skillful in song, but the chatter of house sparrows in a large bay tree outside our kitchen door has resulted in the bush being known as 'Tower Hamlets'. Surely they are passing on the gossip of the day, and in spite of the fact that the birds are classic LBJs – Little Brown Jobs – I am hugely fond of them. A few years ago their nests were predated by magpies, so we put up a row of nest boxes below the eaves of the house and the sparrows are now resident in their own 'terrace'. They are well-named since any crevice in a house – be it under the eaves or holes in brickwork – will be colonised. They have even nested on the 80th floor of the Empire State Building in New York.

The classic 'Cockney Sparrer' might have a different accent in the Bronx; who knows, but here their chirruping is wonderfully cheering. Populations have declined hugely over the years; a state of affairs blamed on everything from changes in farming practices to modern house building, which makes their access more difficult. In the eighteenth and nineteenth centuries they were eaten or simply killed for their predations on crops, but London lost three-quarters of its sparrow population between 1994 and 2000. Now we do our best to ensure their survival, even if they do eat seeds as well as insects.

There are usually about three broods between April and June, each consisting of between three and five pale grey eggs blotched with darker grey. Incubation takes around two weeks and the young will fledge over the next fortnight. A population once estimated at around 12 million breeding pairs in the 1970s has been reduced to around 5.3 million, but the signs are that numbers are building.

# Fruit of the month

## Blackberry

When I was a boy, we'd take one of mum's kitchen bowls with us and go 'blegging' on a Sunday afternoon in September – raiding the hedgerows down by the River Wharfe, and the brambles at the edge of Ilkley Moor, for luscious blackberries. Mum would make one of her blackberry and apple pies and we would feel well rewarded for our thorn-filled fingers, stained with purple juice. Picked before Michaelmas (see *Feast days for gardeners* at the beginning of this chapter), they always seemed fat and tasty, but much depends on weather conditions where the plants are growing, and if they are starved of water the yield can be disappointing in terms of size and flavour.

By all means go blackberrying – taking advantage of nature's bounty – but grow a blackberry of your own – 'Oregon Thornless' for comfort. Plant it in well-manured soil and train it over a post and wire fence at one end of your veg patch (my grandfather trained his over an old brass bedstead, but nowadays most of these have gone to auction). In pies and crumbles, blackberries are a real late summer treat.

# Fruit from the garden

Apples, apricots, peaches and nectarines, blackberries, blackcurrants, Morello cherries, figs, grapes, melons, pears, plums, autumn-fruiting raspberries.

# Something to read . . .

There are many books about Britain's most revered garden, Sissinghurst in Kent, but that penned by Adam Nicolson – the grandson of Vita Sackville-West and Harold Nicolson – is rather different. Like his grandparents before him, Adam is a highly regarded writer, and the combination of that prodigious talent and his relationship with the creators of the garden gives this book a head start on its rivals. The garden's fame began to burgeon in the 1950s, but Adam goes back centuries to discover the history of the castle and its environs.

The result is a far richer exploration of the innate magic of this place from the earliest days up to the present – a road that is frequently rocky and fraught with peril as well as yielding unexpected delights. Settle down with *Sissinghurst: An Unfinished History* by Adam Nicolson and lose yourself in its atmospheric unraveling of a remarkable story.

# Famous gardener of the month
## Humphry Repton (1752–1818)

Regarded by some as the successor to Lancelot 'Capability' Brown, Repton's style could not be more different. While Brown swept away the formal gardens that surrounded the houses of his clients – bringing the 'countryside' right up to the front door – Repton reinstated 'the picturesque', which allowed for terraces and balustrades along with more intricate features, while still making use of Brown's landscapes and 'borrowed views'. Repton created 'Red Books' for his clients, showing them – by virtue of cut-out overlays – the way in which a particular view could be transformed.

While Brown drew up plans for the estates that he was charged with improving, and then brought in an army of labourers to undertake the work, Repton contented himself with suggesting how layouts might be improved and then leaving the client to arrange the work. It explains why Brown was the equivalent of a multi-millionaire when he died, while Repton was only moderately well off. But he did coin the term 'landscape gardener', which remains in use today. His work can still be seen in such locations as diverse as Woburn Abbey in Bedfordshire and Endsleigh in Devon.

# Wild flower of the month

## Knapweed (*Centaurea nigra*)

At a distance you might think the knapweed is a thistle, but on taking a closer look you will see that the
leaves are not at all spiky and uncomfortable to the
touch, instead they are softer and more linear. The
florets themselves – purple or pink and very occasionally white – push out of a rounded knob of bristly
scales and are beloved of bees, butterflies and assorted
beetles. You'll find knapweed growing to a height of
almost a metre in meadows and on roadside verges,
flowering from June/July right through to September.
The greater knapweed (*Centaurea scabiosa*) is found in
areas of chalk and limestone and its florets spread
wider, giving it more the appearance of a cornflower
than a thistle.

In days of old, young ladies would pull out the
colourful florets of a knapweed flower and place the
knob-like part close to their bosom. If the man of their
dreams were to approach during the day, they would
later check the flower to see if more florets had appeared
as an indication that they had, indeed, found their true
love. Due to the rise in online dating this method of
divination has fallen out of favour, but the tradition is
alluded to in some of the local names below.

*Other local common names*: Bachelor's buttons, Black
soap, Bullheads, Button-weed, Chimneysweep's
brushes, Drumsticks, Hackymore, Hardheads, Horse

knobs, Lady's Balls, Lady's cushion, Matfellon, Shaving brush, Tarbottle, Top-knot.

# Guernsey lilies

The problem with *Nerine sarniensis* is knowing how to pronounce its name. I've never been comfortable with ne-*reen*, which sounds a bit like Doreen. Having undertaken my horticultural education at Kew Gardens, it seemed to me that its Latin name ought to be pronounced nee-*ry*-nee, which is what I have always done. But whatever your chosen pronunciation, do grow a few in well-drained soil at the front of a sunny border, preferably (but not essentially) at the foot of a south-facing wall. The tops of the bulbs need to protrude a little from the surface of the soil, and the clumps can remain undisturbed for years.

On no account allow the foliage of nearby plants to flop over them or they will sulk and fizzle out. When they are happy, they will flower their socks off each September and October, producing 60-cm high stems that carry a glorious umbel of firework flowers in vibrant pink. They last well when cut and are ample compensation for summer's end.

# Things you can do . . .

— Plant daffodil and narcissus bulbs and order tulips.
— Make new lawns from seed (this is the best month to do so).
— Plant hyacinths in bowls for Christmas.
— Plant spring cabbages.
— Pick early apples and pears.
— Plant anemones and freesias in pots.
— Pot up hippeastrums for autumn and winter flowers.
— Cut out a few old stems from rambling roses and tie in new ones.
— Harvest vegetables regularly.
— Plant evergreen shrubs and transplant those that need moving.
— Pot up rooted cuttings.
— Pick all fruits from outdoor tomatoes and ripen them indoors. (Place them in a drawer with a ripe banana to hasten ripening.)
— Dig up and store potatoes in thick paper sacks, keeping them cool and dark.
— Clip yew hedges and topiary.
— Bring tender plants back into the greenhouse before the frosts.

# Things not to do . . .

— Don't imagine that the garden is finished with for this year.
   There are more delights to come, and if there are not, a
   visit to the local nursery or garden centre will provide
   plenty of inspiration. Take a trip there every month and
   buy something in flower – that way you are assured there
   is always something in flower in your garden. Obvious?
   Yes, but neat . . .

— Don't stop mowing. The grass will continue to grow and
   weekly mowing will keep it thick and healthy.

October

Nature now spreads around, in dreary hue,
A pall to cover all that summer knew;
Yet, in the poet's solitary way,
Some pleasing objects for his praise delay;

JOHN CLARE, *The Shepherd's Calendar* (October), 1827

John Clare shares our sadness at the waning of the summer, but urges us to look around for delights that are still to be found. Bright berries, late-flowering perennials, the colouring of foliage on deciduous trees and shrubs, the conkers that will amuse our children and grandchildren, and without autumn, where would be the freshness of spring? The great thing about a temperate climate, as opposed to the sameness of the tropics – day in, day out – is its seasonality that results in an ever-changing scene. This time of year always strikes me as nature's last 'hurrah!'. Leaves are shed and we sweep them up and compost them with a view of returning them to the soil.

October is a reminder of the cyclical nature of the seasons in Britain and of the nature of a gardener's year. With every new one that approaches, we have a chance to do things better. 'Weep if you must, parting is hell; but life goes on, so sing as well,' wrote Joyce Grenfell. I like to think that she and John Clare would have got on famously.

# Weather

Well, it's autumn isn't it? And you know what that means – wind. Probably just after you have swept those leaves into a nice little pile. Then there is the rain. And the frosts that will soon become more frequent, except in the unseasonably mild spells, which occur to confuse us and give us more reason to worry about global warming and climate change. Let's acknowledge those two and park them for the moment. No point in getting depressed. Remember, we're talking about *weather*, not *climate*. The days are significantly shorter now, and that means the sun is lower in the sky and temperatures are unlikely to see you wearing shorts. Unless you are a postman.

# Day length (Skipton)

|  | 1 October | 31 October |
|---|---|---|
| **Dawn** | 06.35 | 06.29 |
| **Sunrise** | 07.08 | 07.04 |
| **Sunset** | 18.47 | 16.38 |
| **Dusk** | 19.20 | 17.13 |
| **Daylight hours** | 11 hours 33 minutes | 9 hours 28 minutes |

**Clocks go back 1 hour during the early hours of the last Sunday in October.**

# Feast days for gardeners

**21 October**  **Apple Day** – a time to celebrate the apple and all its unusual varieties. A good day to plant one, too.

**Last Sunday in October**  **British Summer Time** ends. The clocks go back in the early hours of the morning and you get an extra hour in bed, but it is dark earlier in the evening from now on and lighter in the mornings – for a while . . .

**31 october**  **Hallow'een** – when witches and . . . well, you know. But it's a great time to make the most of pumpkins and over-sized turnips and swedes by turning them into lanterns.

# Tree of the month

## Japanese maple (*Acer palmatum*)

The Japanese maple has so many different varieties that it becomes addictively collectable. These small trees are so useful where space is limited, provided you give them the conditions they appreciate: shelter from strong winds and scorching sun, both of which cause the foliage to become browned at the leaf tips. In dappled shade and decent soil, they will delight you with their finely cut leaves and elegant habit – some of them upright, others forming low-growing parasols.

There are varieties with purple foliage – 'Red Pygmy' is especially fine – and greenish-yellow leaves tipped with red – 'Tsumo Gaki'. The variety 'Bi Ho' has wonderful orange-yellow stems that show up well in winter (a change from 'Sango Kaku') and 'Aoyagi' makes a statuesque tree with green leaves and green stems – more spectacular than it sounds! The best pale green form is still *Acer shirasawanum* 'Aureum' (which used to be known as *A. japonicum* 'Aureum'), whose pleated leaves are wondrously luminous on unfurling in April.

Their autumn colour at this time of year is a final bonus in the season – anything from butter yellow to fiery scarlet. You can grow them in a pot if you are generous with food and water. I had one that lasted in a container for twenty years.

Height and spread: from 60cm to 5m.

# Music to listen to . . .

I suppose the obvious soundtrack would be *Autumn* from *The Four Seasons* by Vivaldi, but when you've listened to that (for I will not disparage it on the grounds of its popularity), try *Autumn* from *The Four Seasons* by Alexander Glasunov. It's a meaty, triumphal piece that will lift your spirits. I suppose there is a sort of valedictory sentiment in the music of autumn, but if you want to avoid the maudlin and lift your spirits, Leonard Bernstein's Overture to *Candide* will fit the bill. Jauntiness? Try Percy Grainger's *Mock Morris* or *Handel in the Strand*.

So-called 'light music' is often disparaged, but when the weight of life (or of an autumn of leaf-sweeping) dampens your spirits, the works of Ronald Binge – *The Watermill* – and Eric Coates – the *Westminster Waltz* and *Knightsbridge March* – will perk you up. Two conductors, Ronald Corp and Iain Sutherland, have been champions of light music and their albums are well worth a listen.

## Vegetables from the garden this month

Jerusalem artichokes, broccoli, cabbage, carrots, cauliflower, celeriac, celery, endive, kohl rabi, leeks, lettuce, parsnips, peas, spinach, swedes, turnips.

## Vegetables in store

Beetroot, carrots, onions, potatoes, shallots.

## Something to muse upon . . .

One of my own monthly delights – sometimes we go longer than a month – is a phone call with Alan Bennett.

I remain flattered that he values our conversations, which are usually of a prosaic and down-to-earth nature, but always sprinkled with snippets of gossip or anecdotes that both of us prize. I enjoy making him laugh and value his books, particularly the diaries: *Keeping On Keeping On, Telling Stories* and *Writing Home*. We meet up with him and his partner Rupert from time to time and with typical generosity of spirit he came and gave *An Evening with . . .* in our barn when we turned it into a tiny theatre.

At the very end of his play *The History Boys*, the teacher, Hector (played in the original production by Richard Griffiths) explains to his pupils the importance of sharing what they have learned; what they have experienced of life. It is a simple exhortation, but one which I have always found quite moving since it equates with my own evangelical approach to defining the value of gardens and gardening and their ability to enrich our lives:

Hector:
Pass the parcel. That's sometimes all you can do. Take it, feel it and pass it on. Not for me, not for you but for someone, somewhere, one day. Pass it on, boys. That's the game I wanted you to learn. Pass it on.

ALAN BENNETT, *The History Boys*, 2006

# Flower of the month

## Aster

The words 'Michaelmas daisies' probably conjure up images of dreary, dust-laden flowers atop stems of browning leaves that occur in abundance along railway embankments. That might once have been the case – and those very plants still hold sway on branch lines – but modern varieties are brighter of hue, healthier of constitution, and a valuable addition to beds and borders in October when most other flowers are packing up and going home. The flowers now may be lavender blue, zingy magenta, rosy pink, luscious lilac or pure white, and they are as good for cutting as they are for leaving in the garden to cheer up grey autumn days.

Most grow between 60cm and 1m tall. Seek out the snappily named 'Andenken an Alma Potschke', which is vibrant magenta, 'Purple Dome' and 'Violetta', whose colour is obvious, and the soft pink 'Patricia Ballard'. The best ones are classified as *novae-belgii* and *novae-angliae* . . . er . . . asters, in spite of the fact that unhelpful botanists have now decided that they should be called Symphyotrichum rather than aster. Their only drawback is that they do benefit from being divided up and replanted every three years – they spread and die out in patches over the years – but at least it is a chance to share your favourites with friends.

# Flowers in the garden

Abelia, Aconitum, Japanese anemones, Asters (symphyotrichum), Arbutus, *Buddleja asiatica*, Calluna (heather), Ceratostigma, Chrysanthemums, Clematis, Colchicum (autumn crocus), Crocus (autumn-flowering varieties), *Cyclamen hederifolium*, Elaeagnus, Erica, Fatsia, Gaillardia, Gentiana, Geum, Hebe, Helianthus (annual and perennial sunflowers), *Hydrangea paniculata*, Hypericum, Kniphofia (Red-hot poker), Liriope (lily turf), Lithodora, Nerine (Guernsey lily), Osmanthus, Perovskia, Physalis (Chinese lantern flower), Scabiosa (scabious), Schizostylis (kaffir lily), Sedum (ice plant), Solidago (golden rod), Teucrium, Verbena, Zauschneria.

# Ornamental fruiting trees and shrubs

Berberis (barberry), Callicarpa (beauty berry), Celastrus, Cotoneaster, Crataegus (hawthorn), Euonymus (spindle tree), Hippophae (sea buckthorn), Malus (crab apples), Pernettya, Pyracantha (firethorn), Roses (hips), Sambucus (elder), Skimmia, Sorbus (mountain ash, Whitebeam), Viburnum, Vitis (vines).

# A garden to visit in October
## Westonbirt Arboretum

October marks the onset of autumn colour, which, with any luck – and no following wind – will last into November. Arboreta (collections of trees) are great places to visit at this time of year. They offer hope in spring when their leaves are unfurling, and many of them put on a spectacular grand finale in autumn. Maples, liquidambar (sweet gum), stag's horn sumach (*Rhus typhina*), deciduous euonymus, beech trees and mountain ash are only some that have a glorious final 'Hurrah!' before their leaves fall and their branches remain bare through the winter. Westonbirt, in Gloucestershire, is one of the best places to see the spectacle, but look up woodland gardens in your own area and savour the arboricultural curtain call.

# Hyacinths

Home-grown Christmas presents are never less than well received, and a few hyacinth bulbs, potted up now with their 'noses' just showing and stood in a cool, dark place (a cellar or windowless shed) until December (they need eight to ten weeks in the dark to establish their roots), will have produced fat and promising flower buds in time for the festive season. Use 'prepared' bulbs, which are quicker to flower, and pot them up in peat-free multipurpose compost in bowls or pots with holes in to ensure good drainage. Space the bulbs just 5cm apart. Don't mix up the colours – they will flower at slightly different times and make for a less spectacular show.

Check that they do not go short of water in their dark dormitory and bring them indoors when the flower spike can be seen to have pushed up above the 'neck' of the bulb. The cooler you keep the bulbs, the longer will their flowers – and their wonderful fragrance – last. Should the flower spikes look like toppling, make sure they have really bright light, and push a short length of stout wire down through the flower stalk and into the bulb. It may seem cruel, but since the bulbs are generally discarded after flowering that really doesn't matter.

Oh, you plant them in the garden after flowering? Mmm. Well, you do run the risk of beds and borders peppered with flowerless and unidentifiable bulbs in future years, but even if you do use the wire method, the chances are that the bulbs will survive if you remove the wires as the flowers fade.

# Bird of the month

## Magpie (*Pica pica*)

Do you remember so many magpies around when you were a child? No, neither do I, but they are now ubiquitous. The Victorian gamekeepers persecuted them on account of their predations on young pheasants, partridges and grouse, and it probably took them a century to increase their numbers. I did wonder whether I should encourage them by giving them pride of place this month, but they are with us whether we like it or not, in spite of their unfortunate predatory habits on other birds' nestlings and eggs. It is perhaps best to think of them as 'nature's undertakers', feeding on the carrion that results due to encounters with cars.

So much country lore surrounds them and I confess to still saluting when I catch sight of 'One for sorrow, two for joy . . .', so well known not to need re-telling, but I do like the gentler version discovered in Ireland by those guardians of the nursery rhyme Iona and Peter Opie:

One for sorrow, two for joy,
Three for a kiss and four for a boy,
Five for silver, six for gold,
Seven for a secret never to be told,
Eight for a letter from over the sea,
Nine for a lover as true as can be.

The chattering, clattering call of the magpie is a familiar sound, but their reputation for thieving, perpetuated by Rossini, would seem to be undeserved. That said, the collective noun for a group of magpies is 'a mischief'. Their nest is a dome-shaped affair of twigs in trees and bushes where between five and eight eggs, pale blue, blotched with olive brown, are laid in a single brood in April or May. They hatch in two-and-a-half to three weeks and the young take almost a month to fly. Aside from carrion, eggs and nestlings, the birds also feed on insects and seeds. Not all magpies breed – nature's way of population control you might argue – but the UK has around 600,000 breeding pairs, and most of them are territorial, travelling only within about 12 acres.

# Autumn colour

This is the month when it starts – that blushing of the countryside when the green chlorophyll in the leaves breaks down and the brighter anthocyanins are revealed. Not all deciduous plants turn brilliant shades of red and yellow – many simply turn brown before they fall off – but those that do are rightly prized for their late season spectacle. What makes for a good autumn where foliage colour is concerned? There are countless theories, but it seems to me that relatively dry soil, steadily lowering temperatures (rather than a sudden drop, which can result in leaves being shed prematurely) and – obviously – a lack of strong winds that can denude the trees with rapidity, will produce the most spectacular display.

Among the trees and shrubs that colour best are the maples, deciduous viburnums, sumach, liquidambar, hamamelis (witch hazel), which colours up to match its flowers – yellow-flowered varieties have yellow autumn tints, red-flowered varieties are more ruddy. The leaves of birches turn butter yellow and those of the sorbus (rowan) turn red. Cornus species (dogwoods) are good, too, as is amelanchier (snowy mespilus). *Nyssa sylvatica* (black tupelo) has the added advantage that its rich red colouring is partnered by a glossy surface. Observe it against a blue sky on a sunny day and you – if not the leaves – will be blown away.

# Fruit from the garden

Apples, figs, grapes, medlars, pears, quinces, autumn-fruiting raspberries.

# Fruit of the month

## Grapes

Should you own a stately home, which possesses a walled kitchen garden with a range of lean-to greenhouses, each devoted to a different crop – peaches and nectarines, camellias, orchids – then you will probably have your own vine. If you do, it will most certainly be of the best variety of all: 'Muscat of Alexandria'. Ah, the muscat! No grape has such a delicate and intoxicating perfume; no grape has such a delicious flavour; no grape is so temperamental and tricky to grow, but when it is grown well . . .

Best perhaps to stick to the tried and tested 'Black Hamburgh', which is rather easier to cultivate. True, its skin may be a little more robust than that of 'Muscat of Alexandria', but it is still a valued grape variety and once you have mastered the techniques of pruning and training, thinning the bunches of fruit and ventilating wisely to prevent outbreaks of mildew, you will understand why Lancelot 'Capability' Brown planted the one at Hampton Court which still thrives.

Should you wish to grow a grape under glass, read up about it and learn the technique, then it can become a challenge worth undertaking. Alternatively, in the southern half of the country, you can grow grapes such as 'Phoenix' or 'Pinot Noir' outdoors on a post and wire framework and make your own wine. A friend did just that. When a dinner guest sampled the vintage and asked where the grapes had been grown, his host pointed to the vines just outside the window. 'Hmmm,' remarked the guest. 'Doesn't travel well, does it?'

I am sure you will do better.

# Something to read . . .

Every so often a film producer will turn to a delightful short novel by James Hilton – *Goodbye Mr Chips* – the story of a schoolmaster whose long tenure at Brookfield, a boarding school for boys, was marked by his fondness for Latin and for the young in his care who became, for him, like family. Martin Clunes gave a delightful portrayal on television a few years ago, and I have a shameless fondness for the musical film, made in 1969, which starred Peter O'Toole in the leading role and Petula Clark as his wife.

I re-read the book occasionally and disappear into a world I never knew, as a pupil at a Yorkshire secondary modern state school in the early 1960s. This is the opening of the book, which first saw the light of day in 1934:

When you are getting on in years (but not ill, of course), you get very sleepy at times, and the hours seem to pass like lazy cattle moving across a landscape. It was like that for Chips as the autumn term progressed and the days shortened till it was actually dark enough to light the gas before call-over. For Chips, like some old sea-captain, still measured time by the signals of the past; and well he might, for he lived at Mrs Wickett's, just across the road from the school. He had been there for more than a decade, ever since he finally gave up his mastership; and it was Brookfield far more than Greenwich time that both he and his landlady kept.

'Mrs Wickett,' Chips would sing out, in that jerky, high-pitched voice that had still a good deal of sprightliness in it, 'you might bring me a cup of tea before prep, will you?'

When you are getting on in years, it is nice to sit by the fire and drink a cup of tea and listen to the school bell sounding dinner, call-over, prep, and lights out. Chips always wound up the clock after that last bell; then he put the wire guard in front of the fire, turned out the gas and carried a detective novel to bed. Rarely did he read more than a page of it before sleep came swiftly and peacefully, more like a mystic intensifying of perception than any changeful entrance to another world. For his days and nights were equally full of dreaming.

JAMES HILTON, *Goodbye Mr Chips*, 1934

You see why I recommend it for autumn? It sounds like a sleepy sort of tale. It is wistful, certainly, but it is also a story of touching sentiment and surprising emotion as we learn of the earlier life of Mr Chipping and his arrival at Brookfield in 1870, of him falling in love and marrying, which nobody ever thought he would do, and of triumph and tragedy in equal measure. It is a story with a good heart, and we all need one of those from time to time.

Should you prefer a gardening book, then seek out *Thoughtful Gardening* by Robin Lane-Fox – as readable as he is wise.

# Famous gardener of the month

## Sarah Raven (1963–)

Matchless of pedigree (Sarah's father was John Raven who, as well as being a classics don at King's College, Cambridge, wrote the eminently readable *The Botanist's Garden*), she is married to the Sissinghurst historian Adam Nicolson. Sarah is an accomplished gardener and garden writer. She also runs gardening courses at her small farm, Perch Hill in Sussex. Originally trained as a doctor, Sarah's practical gardening knowledge is prodigious and her stylish approach to planting is inspiring.

Seek out and enjoy her beautifully illustrated book *Wild Flowers* – a wondrous celebration of the British flora. As well as writing *The Cutting Garden*, Sarah also runs a mail-order company specialising in bulbs, seeds and flowers that can be grown for cutting, along with classy garden sundries. Her enthusiasm is infectious and her commitment to good gardening techniques is demonstrated by her ability to grow all kinds of plants really well.

# Rewilding

As someone who has created from seed a glorious 2-acre wild flower meadow, I have mixed feelings about the great movement towards 'rewilding', which does not, as some might assume, mean letting land lie fallow. Rewilding takes careful management (and in some instances the introduction of long-horn cattle) and is not simply a matter of turning one's back on the land and letting nature take over. God can do wonderful things in a garden, but when he has it entirely to himself it ceases to be a garden and becomes a patch of overgrown scrub, frequently replete with empty crisp packets, beer cans and dog poo bags lobbed there by uncaring souls who imagine that it is a rubbish dump.

There is a place for wilder areas in any garden – where birds, bees and butterflies are catered for. Funnily enough, birds, bees and butterflies seem to like gardens filled with exotic flowers, too, if my own patch is

anything to go by. The answer is, of course, balance.
Have a patch of grass that you do not mow, except to
cut off the 'hay' in late summer, but enjoy your striped
lawn, too, as will the blackbirds who find it harder to
locate worms to feed their young in long grass. I deplore
those who endeavour to make gardeners feel guilty for
growing bright hardy perennials, roses and summer
bedding plants, in the belief that if they grow only
native flowers they will be doing nature a greater
service. It ain't necessarily so.

The greatest enemy of nature (and the greatest aid to
surface flooding) is the block-paved front drive where
the electric car (in the interests of not exacerbating
global warming) is parked. Run your existing car into
the ground before you buy another (far more environmentally friendly), and plant low-growing shrublets,
ground cover and carpeting herbs around the parking
area to ameliorate flooding and to be more beneficial to
wildlife. My garden teems with all manner of insects,
birds and animal life, but it is a garden not a wilderness,
and I refuse to feel guilty for growing beautiful plants in
a design that I find appealing. I love my wild flowers,
but I love my beds and borders, too. So there . . .

# Wild flower of the month

## Blackthorn (*Prunus spinosa*)

Hard of wood (it is used to make walking sticks) and
strong of thorn (reputed to be able to puncture tractor

tyres), the blackthorn is one of the delights of the spring hedgerow when its white blossom turns them into anchored cumulus clouds. Then, in autumn, come the round, black sloes, prized for making sloe gin – best, it is said, when they have been frosted. If you are making a dog-proof, stock-proof hedge, the blackthorn is really useful since it suckers widely and rapidly makes a hedge impenetrable. The leaves are small, oval and green, and the term 'blackthorn winter' has come to mean a cold snap in spring when the temperature is so low that the blossom could be mistaken for snow. (For what to do with your sloes, see *Fruit of the month* for November.)

*Other local common names*: Black-haw, Bullen, Bullison, Bullister, Egg-peg-bush, Pig-in-the-hedge, Scrogg, Slaathorn, Slacen-bush, Slon-tree, Snag-bush.

*For the fruits*: Bullens, Heg-pegs, Hedge-picks, Slags, Snags, Winter Kecksies, Winter-picks.

# Things you can do . . .

— Towards the end of the month, begin planting bare-root trees and shrubs dug up from open ground.
— Sweep up fallen leaves regularly.
— Make new lawns from turf.
— Make new flowerbeds and borders, digging in plenty of well-rotted compost and manure.
— Spike the lawn to help surface drainage and rake out dead grass (thatch) and moss.
— Pull out summer bedding and plant spring bedding such as wallflowers, pansies, sweet Williams.
— Plant spring-flowering bulbs such as daffodils, narcissi, crocuses and dwarf irises.
— Dig up and store dahlias when they have been frosted, or chance leaving them where they are and mulch them with well-rotted compost or bark to insulate them from frost.
— Plant border perennials on light, well-drained soils.
— Take hardwood cuttings of shrubs and root them outdoors in small trenches on the veg patch.
— Plant spring cabbages.
— Dig up and store gladioli.
— Clean greenhouse glass to allow maximum light penetration in winter.
— Put up nest-boxes to allow birds roosting sites for the winter.
— Make sure feeders are topped up and that the birds have fresh water.

# Things you should not do . . .

— Avoid planting new beds and borders if your soil is heavy clay. You'll be better off waiting until spring when there is less risk of the plants rotting off.
— Don't wait until all the leaves are down before you sweep them up. Left too long on the lawn they will kill out the grass and they may also encourage rotting of herbaceous plants.
— Don't keep walking the same path on a soggy lawn – you'll turn it to mud.
— Don't forget to buy spring-flowering bulbs.
— Don't be too tidy – birds and insects will enjoy seedheads that are ripening.

November

The great thing about the garden in November
is that no one expects anything of it.

CHRISTOPHER LLOYD in conversation with Alan Titchmarsh

Keeping our spirits up in November is hard. We know what's ahead: shortening days and lowering temperatures. I always think of November as a 'dust-pan and brush' month – when the garden starts to decay and many gardeners either become obsessed with tidying it up or else turn their backs and let nature get on with things – emerging again in spring to take stock and start the new season. Good gardeners, of course, are out there in all weathers; well, at least until our feet freeze or the rain runs down the back of our neck.

When it comes to orderliness, most of us try to strike a balance between being overly assiduous and too *laissez-faire*, but as one who is naturally tidy-minded indoors and out, the neatness of beds and borders – at least around the edges – makes up in some way for nature's most disorderly time of year.

# Weather

November is likely to be one of the wettest months, though hopefully not as wet as the autumn of 2000, when 503mm of rain (that's almost 50cm) fell during the months of September, October and November, making it the wettest autumn since records began in 1766. Being out in all weathers might be a source of pride to some, but walking on wet or frozen soil does more harm than good – destroying its structure and impeding drainage, so you have a legitimate excuse for staying off wet or freezing earth.

Ah, yes. Freezing. When I was a boy growing up in Yorkshire in the 1950s, it seemed to me that frosts regularly arrived in September, blackening the dahlias and shortening the summer. Now it is often November before temperatures plummet, but plummet they will – gently one hopes, to give plants the chance to slip quietly into dormancy rather than shedding their leaves in shock. Still, I cherish the banner that regularly adorns the fence surrounding our local purveyor of garden buildings at this time of year: NOW IS THE WINTER OF OUR DISCOUNT SHEDS.

# Day length (Skipton)

|  | 1 November | 30 November |
|---|---|---|
| Dawn | 06.31 | 07.19 |
| Sunrise | 07.06 | 07.59 |
| Sunset | 16.36 | 15.53 |
| Dusk | 17.12 | 16.33 |
| Daylight hours | 9 hours 24 minutes | 7 hours 48 minutes |

# Feast days for gardeners

**5 November**    **Guy Fawkes' Night** – a time to celebrate the failure of a plot to blow up the Houses of Parliament. Held outdoors in a part of the garden where the remains of a bonfire will not be offensive.

**11 November**    **Armistice Day/Remembrance Day** – Poppy Day, commemorating those who lost their lives in conflict, especially the two world wars. Poppies grew on Flanders Field and have been used as a symbol of remembrance since 1918.

| | |
|---|---|
| **11 November** | **Martinmas** – a celebration of the life of St Martin, whose compassion towards beggars and the destitute is remembered on the same day as the Fallen. |
| **30 November** | **St Andrew's Day** – the patron saint of Scotland, whose commemorative flower is the Scottish thistle. |
| **Advent Sunday** | falls towards the end of this month, commemorating the coming of Christ in advance of the celebration of his birth at Christmas. |

# Tree of the month

## Strawberry tree (*Arbutus unedo*)

Sometimes known as the Killarney strawberry tree, arbutus gets its name from the colour and texture of the spherical fruits, which decorate the branches after the clusters of white or pale pink urn-shaped flowers have faded, though flowers and fruits are carried at the same time in autumn – the fruit taking a year to come to maturity. It is an evergreen that is found in the wild from south-western Ireland right across to the Mediterranean. Bear in mind that arbutus enjoys most the milder, damper countries, where it will make an impressive round-headed tree and at maturity it has wonderfully attractive peeling bark, russety in colour.

It will grow in any well-drained soil, whether acid or alkaline. The fruits themselves are edible, but are most frequently made into jams. Owing to the fact that it has green leaves, white flowers and red fruits, arbutus is the national tree of Italy. It is popular in the western states of the USA and I wish it were more widely grown here in the UK.

Height and spread: 8m, but there are lower-growing forms such as 'Compacta', which makes a decent-sized shrub rather than a tree.

# Music to listen to . . .

At the request of HRH The Prince of Wales, Patrick Hawes composed *The Highgrove Suite* in 2010 and it is a charming collection of pieces, each inspired by Prince's now famous garden in Gloucestershire. It would be all too easy to suggest more well-known pastoral pieces for autumn (I'm avoiding Vivaldi's *Four Seasons* or it will crop up every quarter), but I'll stick with this and Patrick Hawes' *Lady Radnor Suite* which has a similar feel, and both are available on the same album. Oh, and one more composer who deserves to be rescued from relative obscurity: Christopher Ball, who died in April 2022, but leaves behind some glorious arrangements and compositions. Seek out his Oboe and Horn Concertos and admire their freshness and verve.

# Vegetables from the garden

Jerusalem artichoke, Brussels sprouts, cabbages, cauliflower, celeriac, celery, endive, kohl rabi, leeks, parsnips, savoys, spinach, swede.

# Vegetables in store

Beetroot, carrots, onions, parsnips, potatoes, shallots, swedes, turnips.

# Something to muse upon . . .

I am not a pessimist by nature, so this poem by Thomas Hood seems to me not only inaccurate but also rather defeatist. But I include it so that in those moments of autumnal dejection you do not feel alone.

## No!

No sun – no moon!
No morn – no noon –
No dawn –
No sky – no earthly view –
No distance looking blue –
No road – no street – no 't'other side the way' –
No end to any Row –
No indications where the Crescents go –
No top to any steeple –
No recognitions of familiar people –
No courtesies for showing 'em –
No knowing 'em!
No travelling at all – no locomotion –
No inkling of the way – no notion –
'No go' – by land or ocean –
No mail – no post –
No news from any foreign coast –
No park – no ring – no afternoon gentility –
No company – no nobility –
No warmth, no cheerfulness, no healthful ease,
No comfortable feel in any member –
No shade, no shine, no butterflies, no bees,
No fruits, no flowers, no leaves, no birds,
November!

THOMAS HOOD, *No!*, 1844

# Flower of the month

## Lily turf (*Liriope muscari*)

This is one of those plants that looks insignificant for most of the year but which, come autumn, erupts with small purple flowers that are carried on slender spikes that tower over the low-growing fountains of glossy evergreen leaves. Its main use is as ground-cover in any ordinary soil in sun, dappled or quite heavy shade, though it will flower best where it has a reasonable amount of light. With liriope timing is everything, for it comes into its own when most self-respecting border flowers are getting ready for bed, and it is all the more appreciated for its late arrival.

It looks at its best when planted *en masse*, when it will form a thick rug that sings with subtle colouration from late summer to autumn, and it is easy to propagate by division come early spring. Celebrate it for its late arrival and use it where other plants have struggled to get a hold. It will seldom let you down.

# Flowers in the garden

Arbutus (strawberry tree), Asters (Symphyotrichum, Michaelmas daisies), Colchicum (autumn crocus), Crocus (autumn-flowering varieties), *Cyclamen hederifolium*, Erica (heaths), Fatsia (false castor oil), Gentiana (autumn-flowering gentians), *Hebe* 'Autumn Glory', *Iris*

*unguicularis*, *Jasminum nudiflorum*, Liriope (lily turf),
*Mahonia fortunei*, Physalis (Chinese lantern flower),
*Prunus subhirtella* 'Autumnalis', Schizostylis (kaffir lily),
*Viburnum farreri*, *Viburnum tinus*.

# Ornamental fruiting trees and shrubs

Berberis (barberry), Callicarpa (beauty berry),
Celastrus, Cotoneaster, Crataegus (hawthorn),
Euonymus (spindle tree), Hippophae (sea buckthorn),
Ilex (holly), Pernettya, Pyracantha (firethorn, Roses
(hips), Skimmia, Sorbus (mountain ash and white-
beam), Viburnum, Vitis (vines).

# A garden to visit in November

## Knoll Gardens

Neil Lucas took over Knoll Gardens at Wimborne in
Dorset back in the 1990s and has made it nationally
famous as a garden where ornamental grasses are used
to dramatic effect. The garden was begun in the 1970s,
which accounts for the mature trees it contains, but it
is over the last thirty years that it has become a place
of pilgrimage for those who want to find out how
grasses can be integrated into the modern garden.
Whether your garden is large or small, formal or informal, you will find that a visit to Knoll Gardens at this

time of year emphasises how useful grasses are in prolonging the season of interest – cheering and spectacular when rimed by frost and spot-lit by early morning sun.

Not that grasses are the only attraction at Knoll – at other times of year perennials and shrubs come into their own, and the plant sales area will tempt you at any time of year. Oh, and Neil is a very nice man whose vast knowledge of plants is cheerfully dispensed to anyone sensible enough to ask his advice.

# Bare-root planting

We've all become so used to planting trees and shrubs from containers at whatever time of year we like, we often forget the economies that can be made by planting bare-root specimens which have been dug up from nursery rows with little earth clinging to them during the dormant season. Between the months of November and March, deciduous plants lose their leaves and slip into dormancy, which means that they can be uprooted and moved to a new spot with relatively little shock to their system. Well, to be honest, it must be one hell of a shock to wake up in spring and find that you are somewhere completely different without half the roots you had when you nodded off in autumn. But such plants are eminently capable of recovering, provided the roots are not allowed to dry out during the move.

Unpack the trees or shrubs on arrival and soak their roots in a bucket of water overnight. Heel them in (plant them in a temporary spot) if your planting site is not ready. Before you plant them properly, snip off any long or damaged roots and prepare the soil well – working in a reasonable amount of organic matter (well-rotted garden compost or manure) to give them a head start. On really heavy clay soil, don't work in too much muck or you'll create a sump, which can collect water and result in the plants drowning in wet weather. Plant so that the soil mark on the stem is level with that of their new location and do remember to water them well in dry spells during the first year of establishment.

And a word about size: the larger the plants, the longer they will take to settle in. Plants are like people: babies seldom notice when you move house; OAPs find it more unsettling. The younger the tree, the quicker it will become established. As a rule of thumb, I try to avoid planting trees that are much taller than I am. (Well, much taller than 1.8m, anyway.) Young ones will romp away and often overtake taller specimens that were bought for instant effect. Hedges are most readily planted as bare-root specimens – they are a darned sight less expensive that way, too.

# Bird of the month

## Redwing (*Turdus iliacus*)

At first you might think it is a song thrush; then you notice the russety flank underneath the wing and dark brown stripes on its face that make it look rather more cross than the thrush. It is slightly smaller, too. The redwing is closely related to our resident songster, but it is a winter visitor, arriving from Iceland, Scandinavia and Russia, though a very few breeding pairs do remain in the far north of Scotland where the cooler temperatures suit them. The migrants arrive in mainland Britain from October onwards – survey results vary suggesting anything from 1 million to 8.5 million of them – making the crossing from Iceland (to the northern and western UK) or the Scandinavian coast (to the rest of mainland Britain) overnight.

The night-time migration ensures they are cooler than in the daytime and the risk of predation is reduced. Once here, they feed on worms, berries and insects, the flocks moving on quite quickly to new pastures when it suits them. They are more fearful of human activity than the song thrush and it is harder to get closer to them in my wild flower meadow, where a flock of fifty or more arrives for a short stay each autumn. Their flight is direct, like the song thrush, rather than the series of swoops noticed in their larger relative the fieldfare, with whose flocks they regularly mingle.

When they do breed here, they nest on the ground, perhaps against a tree trunk, and there are generally two broods between April and June, the four to five eggs being pale blue with brown speckles. The eggs hatch in a couple of weeks and the youngsters fledge a fortnight later. On October and November nights you may hear the flocks arriving – their 'tseep', 'tseep' calls glistening like oral silver in the night air.

# Leafmould making

I know it's tedious, but sweeping up fallen leaves in autumn can at least result in rich, brown, crumbly leaf-mould, which makes wonderful soil enrichment. Except that sometimes it seems to take forever. The answer is . . . patience. And the right method. Make a post and wire bin in a corner of the garden. Pile in the leaves and firm them as you do so. (That's important; if there is lots of air between them and the heap dries out, the leaves will stop rotting.) Keep the heap moist – a lump of old carpet thrown over the top will help – and sprinkle a bit of soil in now and again to add more bacteria. Yes, it will take a year for the stuff to break down, but very little labour will have been involved, and you'll have free soil enrichment.

No room for a bin? Try piling them into a black bin liner, which will help retain moisture. Shake it up every month or two and then firm it down again. Personally I'd rather look at a small wire netting cylinder behind the garden shed than a bin liner, but it's up to you. Either way, you'll feel suitably virtuous when you add the friable mixture to your soil during winter cultivations.

# Fruits in store

Apples, medlars, pears, quinces. Raspberries and gooseberries in the freezers.

# Fruit of the month

## Sloe

I made the blackthorn (*Prunus spinosa*), which produces sloes, my 'Wild flower of the month' in October, but the fruits themselves deserve recognition. They are too tart and astringent on the tongue to make them a suitable alternative to blueberries on your breakfast cereal or porridge, but they make a fine restorative drink when it comes to sloe gin. Gather the berries (preferably once they have been frosted we are told); prick them with a needle and place them in a large jar with the gin, shaking the mixture once a day for a week, then leaving it for two or three months before straining it into bottles. The longer you leave it, the better it tastes, they say.

Quantities? Trial and error. Some add sugar to the mix; others prefer the purer method. That much is up

to you. I do add a little sugar myself and am enjoying
sloe gin that we made five years ago. While I would
hardly rank myself as an accomplished sommelier, I
have to say that . . . it has not gone off. One country-
woman tells of an uncle who remarked that he liked his
women fast and his gin sloe. I doubt that he could say
that nowadays.

# Something to read . . .

Autumn and winter are the months for dreaming and
for planning ahead: for making next year's garden even
better than it was this year and for seeking inspiration
when it comes to garden design. We can all try 'do-it-
yourself' garden designing, but taking ideas and inspiration from accomplished designers whose work we
admire is what gardeners have done for centuries. There
may be 'nothing new under the sun', but there are
certainly great designers who can add freshness of touch
to garden layouts and whose ideas can be adapted to our
own situation. Even if no great transformations are
planned, there is pleasure to be had in looking at the
work of others from the comfort of a winter fireside.

One book I especially cherish is Arabella Lennox-
Boyd's *Designing Gardens*. Her experience of different
countries, differing landscapes and the varied requirements of her clients has led Arabella to produce a
mouthwatering testament to her skills and accomplish-
ments. There's a second book, too: *Gardens in My Life*.

Turn the pages of either and escape the murky weather of a November day . . .

# Famous gardener of the month
## Roy Lancaster CBE VMH (1937–)

When it comes to plantsmanship, Roy Lancaster heads the field. After many years of working at Hillier's, the Hampshire tree and shrub nursery famed in the twentieth century for the comprehensiveness of its catalogue – *Hillier's Manual of Trees and Shrubs* – in whose creation Roy was instrumental, Roy's expertise gained as curator of the Hillier Arboretum was spread even more widely as a writer and plant collector; following in the footsteps of his heroes – men like Frank Kingdon-Ward and Ernest 'Chinese' Wilson. His passion began in early youth – in the countryside around Bolton in Lancashire in search of the wild asparagus – and blossomed into plant-collecting trips across the globe, most especially in Nepal and the surrounding regions.

Appearances on BBC's *Gardeners' World* and *Gardeners' Question Time* consolidated his reputation and he remains something of a guru in the world of horticulture. If you find a plant you don't know, 'Ask Roy Lancaster' is a frequent recommendation. His autobiography *My Life With Plants* will tell you more.

# Planting tulips

Spring bulb planting time begins in September, when daffodils and narcissi that have lived in the garden all year round start making their roots. Tulips, though, can wait until November to be planted, which is useful if they are replacing summer bedding that, in our extended summers, can often carry on looking good well into October. Unlike daffs, tulips are not always successful when left in the ground, which is why they are usually dug up after flowering and the larger bulbs from the newly formed cluster are stored for replanting in autumn.

But if you don't want all the faff this entails, plant your tulips bulbs in bed and borders 20cm deep, which will allow them to survive more happily through the summer than if they were planted more shallowly. Not all varieties are suited to this method of cultivation, but I do have several that have done well for me, among them 'Spring Green' and 'Abu Hassan'. If you are prepared to experiment, what have you got to lose? But remember that very few tulips can survive for more than a single season when planted in grass. You'll have to add to them every year and cope with the oldies producing no more than a single paddle-shaped leaf, which is hardly a thing of beauty.

Tulips, for me, are at their most sensational in two narrow borders either side of a path leading up to my greenhouse, and in a dozen large terracotta pots, where

I plant ten of each different variety and brighten up the terrace each spring.

# Wild flower of the month

## Spindle tree (*Euonymus europaeus*)

When the spindle tree produces its fruits you will do a double take. First there is the day-glo pink outer casing, and when the cases burst open they reveal bright orange berries. Coupled with bright red autumn foliage tints, this is a great hedging plant, which comes into its own when its companions in the hedgerow are bare and drear. The leaves are green and oval and the flowers, carried in May and June, each have four narrow greenish white petals – not especially remarkable. But in autumn the tree sheds its anonymity. I say 'tree' in deference to its common name and the fact that it can grow to 8 or 9m, but generally the plant makes a large shrub.

The spindle is native to much of Europe (hence its botanical name) and is especially useful on chalky soils. Several species of moth caterpillars enjoy its leaves, as do those of the holly blue butterfly, and birds such as starlings love the berries. The common name is as a result of its straight stems being useful for making spindles for spinning. Today it is still used to make artists' charcoal.

*Other local common names*: Bitchwood, Cat-tree, Death-alder, Dog-timber, Foulrush, Gatteridge, Gatter-tree, Ivy-flower, Pegwood, Pincushion Shrub, Prickwood, Skewerwood, Skiver, Witchwood.

*And for the berries*: Dog-tooth Berries, Hot Cross Buns, Louse-berries, Pincushions, Popcorns.

# Storing fruit

There are few fruit stores built nowadays; the freezer has seen to that. But there is still a need to store 'keeping' varieties of apple and pear and the best way to do that is in a cool but frost-free shed or garage, where the fruits can be placed in well-spaced trays in which the individual fruits are not touching. Line wooden or plastic trays with newspaper and store only undamaged fruit, checking them every couple of weeks and removing any that show signs of rotting. Pears have a habit of ripening suddenly, so do check them every week. With any luck, some of of your harvest will last through until spring. Only when apples have become soft and 'woolly' will you know that the season of enjoying the fruits of your labours has come to an end.

# Things you can do . . .

— Sow a wild flower lawn – this is the best time, so that repeated freezing and thawing can trigger germination. Sow on bare earth, not among established grass, which will impede germination.

— Tidy up beds and borders – but leave seedheads for birds and overwintering insects.

— Dig or fork over vacant ground in the veg patch and make new beds and borders.

— Shorten the growths on hybrid tea and floribunda roses to reduce wind rock. Cut them back by about 30cm.

— Plant trees and shrubs lifted from open ground.

— Clean greenhouse glass and have a good clear-out inside, chucking anything that is past its best.

— Bend the outer leaves over cauliflowers to protect the curds.

— Prune fruit trees.

— Plant tulips.

— Bring indoors early-flowering bulbs.

— Put rhubarb forcing pots in place over established clumps if you want early pickings.

— Dig up and store parsnips and Jerusalem artichokes to make for easier access in soggy weather.

— Rest and relax . . . the year is winding down and you will have time to catch your breath.

# Things not to do . . .

— Don't worry about a thing. This is your recovery time.

December

The colour of springtime is in the flowers;
The colour of winter is in the imagination.

TERRI GUILLEMETS

But then Terri Guillemets is from Arizona, where presumably they lack the winter floral riches of the UK – from forced hyacinths and narcissi, to holly berries and early snowdrops. We Brits frequently pine for warmer, brighter weather when we reach the end of the year, but we need these darker days and lower temperatures to create the seasonality that – along with expertise garnered over the centuries – makes our gardens the envy of the world. And then there is Christmas, when the gardener can make others smile with gifts of terracotta pots filled with burgeoning bulbs or, when imagination fails, a garden centre gift token.

December is a time for dreaming and for planning ahead, hopefully without panicking. Spring, after all, is at least three and probably four months away. Pull that chair closer to the fire, fill that glass, break out the seed catalogues and doze, dreaming of balmier days in a hammock under the boughs of apple blossom. (I do like to offer a positive way of looking at things . . .)

# Weather

Frankly, your guess is as good as mine. The only certainty is that the days will be at their shortest, which at least gives rise to the knowledge that – imperceptibly at first – from the 21st of the month they will get longer and our lives a touch brighter. But crisp December days are to be treasured, and white Christmases dreamed about by everyone, not just Bing Crosby. There may be power cuts and transport strikes, lashing rain and battleship grey skies, but with any luck our families or friends will be around us on the 25th of the month and something we have grown – potatoes or parsnips, frozen raspberries and apple crumble – will grace the Christmas board and give us the satisfaction of knowing that we have contributed in some small way to the festivities.

What's more, if we have made it clear to our nearest and dearest that almost anything to do with gardening – be it book or implement – is a far more appreciated gift than a pair of socks, there will be smiles of gratitude on Christmas morning.

# Day length (Skipton)

|  | 1 December | 21 December | 31 December |
|---|---|---|---|
| **Dawn** | 07.21 | 07.41 | 07.43 |
| **Sunrise** | 08.01 | 08.22 | 08.24 |

| Sunset | 15.52 | 15.49 | 15.57 |
|---|---|---|---|
| Dusk | 16.32 | 16.31 | 16.38 |
| Daylight hours | 7 hours 45 minutes | 7 hours 22 minutes | 7 hours 27 minutes |

**21 December is the shortest day of the year**

# Feast days for gardeners

| | |
|---|---|
| **1 December** | the start of **Meteorological winter**. As if it hadn't started already . . . |
| **21 December** | **The Winter Solstice** – the start of the astronomical winter and the shortest day. |
| **24 December** | **Christmas Eve** – when 'the stockings are hung by the chimney with care'. |
| **25 December** | **Christmas Day** – may it be all that you wish for. |
| **26 December** | **Boxing Day** – St Stephen's Day, when Christmas 'boxes' were traditionally handed out to staff by a grateful employer. |

| 31 December | **New Year's Eve** – (**Hogmanay** in Scotland). When anyone over the age of forty has an early night. The final day of the year in the Gregorian calendar. |

# Tree of the month

## Holly (*Ilex aquifolium*)

Well, it has to be, doesn't it? From singing 'The Holly and the Ivy' to the wreath on the front door and the sprigs behind the pictures, holly is an intrinsic part of Christmas. Why doesn't every holly carry berries? The answer is down to sex. Most hollies carry male and female flowers on separate plants. Only the female plants carry berries and then only if they have been fertilised by the male. Some hollies are hermaphrodite, having flowers that carry both male and female floral parts and so they are self-fertile. To add to the confusion, the yellow-variegated variety 'Golden King' is female, and 'Golden Queen' is male. I ask you!

The variegated hollies are especially good in the garden, since their brightness lasts all year round. They make great hedging plants on almost any soil and their berries are great for the birds in winter, when there will be a battle to see who can get to them first – you or your avian friends. Holly will grow quite tall – up to 10 or 15m – but it withstands clipping – even into topiary shapes offering year-round structure to more formal parts of the garden.

# Music to listen to . . .

I start December listening to J.S. Bach's *Christmas Oratorio*, since it is appropriate for the season but not *too* Christmassy. (By now anyone who works in a shop with piped music will be sick to the back teeth of festive tunes.) When we do finally get to the 'teens' of December, I am happy to launch myself fully into the festive mood. The albums I come back to year after year are *James Galway's Christmas Carol* – glorious seasonal music for flute – and any that consist of John Rutter's music. Not for nothing is he known as 'Mr Christmas'. If I could have only one piece of festive musical fare it would be Nigel Hess's *Christmas Overture* – beautifully orchestrated, triumphant and uplifting, it almost always moves me to tears.

# Vegetables from the garden this month

Jerusalem artichoke, Brussels sprouts, cauliflower, celeriac, celery, kale, leeks, parsnips, savoys, swedes.

# Vegetables in store

Beetroot, carrots, onions, parsnips, potatoes, shallots, turnips.

# Something to muse upon . . .

In *The Wild Silence* by Raynor Winn, the follow-up to *The Salt Path*, which is the story of the author and her husband's walk around the 1000km of the South West Coast Path, Raynor documents the couple's challenging trek in Iceland, but she also records a visit to the island of Iona off the coast of Mull.

Having been there myself, and experienced the unique and powerful atmosphere of this cradle of British Christianity, I was particularly struck by the following conversation in which a resident of the abbey explains its ability to affect quite deeply those who visit its confines:

'There is no spirituality without God. You won't find a thin place out there; it's here in this building, where humans have worshipped God for centuries. This is where you'll find what you seek.'

'I don't think I'm actually seeking anything. I'm just open to what comes. What's a thin place? Is it geological?' . . .

'A thin place is where man can be close to the other world, to God's realm. It's here.'

The notion of the thin place hadn't entered my head for decades, but suddenly the time we'd spent on the Coast Path shone brightly in my memory. I could feel it now, the weeks of headlands and skies, the night of stars and rain, the smell of the weather as it blew in from the sea. I'd sensed something then, a thinness between the wild world and the human, between freedom and containment. We'd walked along the barrier between those worlds and felt something of our natural state of being. Touched a wild connection with the land and held it in the dust on our hands.

RAYNOR WINN, *The Wild Silence*, 2020

There are times in my own garden, and out in the countryside – walking the hills, journeying with my shepherd's crook along that same coast path where Raynor Winn and her husband Moth experienced a deep connection with the natural world – when that 'thinness' is easy to understand. Unlike the Winns, I do have a Christian faith – a quiet one – but I do not agree with the resident of Iona Abbey that experiencing thinness is reserved only for those who believe in God.

It is there, I think, to be experienced by anyone who opens up their heart and mind to allow a deep-seated connection with the earth. People who can see their way beyond the pressures and strictures of an increasingly technological world and find a deeper, yet simpler kind of oneness with nature. Perhaps gardeners are in some way more open to that. Good gardeners, anyway.

# Flower of the month

## Winter-flowering iris (*Iris unguicularis*)

It sits like an unruly rug of linear leaves for most of the year, and then up through that haystack of foliage in December and January push spears of palest amethyst, which open to reveal the most delicate of lavender blue iris flowers that look as though a puff of wind would destroy them. It used to have the more euphonious name of *Iris stylosa* until the rather more unwieldy moniker was bestowed upon it. Pick the flowers when they are in bud and watch them open over an hour or so in a tiny jar of water indoors. This particular iris demands little except to be left alone. Find it a sunny spot in well-drained soil, preferably in front of a south-facing wall, and it will thrive.

# Flowers in the garden

*Arbutus unedo* (strawberry tree), *Chimonanthus praecox* (winter sweet), *Crocus imperati*, *Crocus laevigatus*, *Erica carnea* and *Erica darleyensis* varieties (winter-flowering heathers), Fatsia (false castor oil), Galanthus (snow-drop), Hamamelis (witch hazel), *Helleborus niger* (Christmas rose), *Iris unguicularis*, *Jasminum nudiflorum* (winter jasmine), *Lonicera fragrantissima*, *Mahonia japonica*, *Prunus subhirtella* 'Autumnalis' (winter-flower-ing cherry), Schizostylis (kaffir lily), Sternbergia, *Viburnum farreri*, *Viburnum tinus*.

# Ornamental fruits in the garden

*Aucuba japonica*, Berberis (barberry), *Celastrus orbiculatus*, Cotoneasters, *Crataegus crus-galli* (cockspur thorn), Hippophae (sea buckthorn), Ilex (holly), Pyracantha (firethorn), Rose hips, Sorbus (mountain ash), Taxus (yew).

# A garden to visit in December

Few gardens have as much to offer in the festive season as Kew Gardens, where greenhouses packed full of temperate and tropical plants are augmented nowadays with spectacular light shows that will entrance the entire family. The Royal Horticultural Society's Gardens at Wisley are similarly illuminated and, further north, Chatsworth in Derbyshire offers a vibrant display.

You could argue that the horticultural equivalents of Blackpool Illuminations are hardly dependent upon plants, but you'd be wrong, for the lights themselves are used to create a magical night-time world that shows off to perfection the statuesque form of trees, shrubs and garden architecture. And anything that makes gardens more inspiring to children and adults alike gets my vote. Look in your local paper for other gardens that extend their season of interest by turning themselves into a winter wonderland.

# The holly and the ivy

''Tis the season to be jolly' and 'tis the season when both these plants come into their own, so it is certainly worth planting them to extend the season of interest in the garden, as well as using them as part of your Christmas decorations. But remember, too, that both holly and ivy have a part to play in a garden that is wild-life friendly. Yes, it might be irritating when the birds get to the holly berries before you do, but the fruits play a vital role in ensuring their survival during the otherwise lean months of the year where food is concerned.

Ivy is a pain when it romps away into the upper branches of a tree, where its weight can ultimately topple its host. But where it can be restricted to the trunk of the tree, or allowed to scramble over a shed or outbuilding, it is a valuable source of food for the Holly Blue butterfly (an oddly amusing state of affairs) and can offer shelter for roosting birds such as wrens during the winter months. Variegated hollies and variegated ivies will both offer a brighter livery than the plain green kinds, and ivy is brilliant ground cover in dry shade beneath trees where it is difficult to establish most other plants.

# Bird of the month

## Robin (*Erithacus rubecula*)

> The robin redbreast and the wren
> Are God almighty's cock and hen.

A stalwart of the Christmas card and a favourite of the
gardener, the European robin is an endearing bird
thanks to its friendly nature and ability to sit and wait
for its snacks, while we obligingly turn the soil and
reveal grubs and earthworms for its delectation and
delight. We know it can be an aggressive and territorial
bird, for we will see one robin facing off another, and we
know that it is the male that has the red breast. What is
less widely known is that as well as our resident birds,
there is an influx of robins from Europe in autumn, so
that friendly chap who perches on your spade handle
may well have journeyed far in search of your
hospitality.

Given the title of our 'National Bird' in 1960, the
robin's reputation as the gardener's friend is well
deserved: the birds feed on all manner of insects as well
as seeds from bird feeders and bird tables. Of all our
garden birds, the robin is the one that can become the
most tame – even eating out of the hands of those who
build up a trusting relationship with it. Nesting in all
kinds of nooks, crannies and crevices, from flowerpots
to old kettles, as well as open-fronted nest-boxes erected
especially for them, robins may have up to three broods

of chicks between April and June. Five or six white eggs speckled with reddish brown are laid and will hatch in two weeks, the young taking a further two weeks to fledge.

The UK population numbers around 4 million.

# Christmas tree

Prince Albert is rightly credited with introducing the Christmas tree into British homes in the first half of the nineteenth century, but the bringing indoors of 'A Paradise Tree' has its origins in Europe several centuries prior to that. Pyramids of sticks were decorated with fruits and sweetmeats to celebrate the forthcoming spring, once the winter solstice was over, as well as 'The Feast of Adam and Eve' on Christmas Eve. The introduction of the first proper 'Christmas Tree' is claimed by both Latvia and Estonia as far back as the fifteenth century, and Martin Luther is sometimes credited with the introduction to our homes of the spruce tree decorated by candles when he was walking in the forest one night and saw the stars twinkling through its evergreen branches in the sixteenth century.

Every year since 1947, Norway has presented Britain with a Norway spruce (*Picea abies*) as a thank you for the assistance given to that country during the Second World War, and while that species reigned supreme in our homes at Christmas for a century, nowadays, in households where pine needles are considered a pain when they entangle themselves with the shagpile, we tend to favour the Nordmann or Caucasian fir, *Abies nordmanniana*, which hangs on to its needles when they dry out, rather than letting them drop. Buy your tree as near to Christmas as you can and saw off the bottom 5cm of stem before placing the tree in a stand with a reservoir that should be kept topped up with water.

Stand it away from any heat source that will exacerbate desiccation, and hope that such precautions will help it last until Twelfth Night. Better still, buy a potted tree and plunge the pot in the garden after the festive season, making sure the compost does not dry out in summer. Bring it indoors each year – repotting it as it grows. A bit of a faff? Yes, but rather satisfying. If you have the space, you can plant out your potted tree and allow it to grow on. I have around twenty years' worth of Christmas trees planted on the edge of our wild flower meadow, ranging in height from 1.2 to 6m. I don't recommend them for small front gardens unless you don't mind living in a cave.

# Fruits in store

Apples, pears, frozen raspberries and gooseberries, jams and marmalade.

# Fruit of the month

## Seville orange (*Citrus x aurantium*)

Alright, so you can't grow citrus fruits – unless you have a greenhouse or conservatory with plenty of space for them in winter – but December is the month to buy Seville oranges to make marmalade that offers us a sweet treat through the darker months of the year. Why Seville oranges? They are more tangy and not so sweet as most dessert oranges, and are consequently higher in pectin, which makes them set better when used in the orange jam we call marmalade. There are apparently around 15,000 orange trees decorating the streets of Seville, and many more in the groves where the fruit is harvested at this time of year.

The fruits generally appear in the shops during January, so I am giving you an early warning! Grab them as soon as you see them, for they quickly run out. There are lots of recipes online, but Mary Berry's is one of the best. Spread some on your toast or make a sandwich and think fondly of Paddington Bear and Her Majesty The Queen.

# Something to read . . .

When our children were growing up I would sit at the bottom of the bed, each Christmas Eve, and read Clement Clarke Moore's *T'was The Night Before*

*Christmas,* written in 1823. Now they read it to their own children and the tradition continues. I realise there is nothing remotely horticultural about the rhyme (except that Father Christmas's nose was described as being 'like a cherry') but traditions are traditions.

To put myself in the Christmas spirit I do read part of Charles Dickens' *The Pickwick Papers* each Yuletide, simply because it conjures up the kind of festive season I suspect all countrymen dream of – devoid of technology and packed with rural riches. What you need is Chapter 28, 'A Good-Humoured Christmas Chapter', which begins: 'As brisk as bees, if not altogether as light as fairies did the four Pickwickians assemble on the morning of the twenty-second day of December . . .' You'll be transported to the top of the Muggleton coach as the Pickwickians make their way to Dingley Dell.

And if you must read about gardening? *Winter Gardens* by Cedric Pollet has the most breathtaking photographic illustrations of plants that have brilliant bark and berries, foliage and flowers. Turn its pages and drool, before making out your order for plants to cheer your life on the darkest days of the year. And as a Christmas present? I blush to recommend *My Secret Garden.* But then if you want to know what my own garden looks like throughout the seasons, that's the way to find out.

# Famous gardener of the month
## Percy Thrower MBE VMH (1913–1988)

I could not end the year without a nod to the man who was my boyhood hero. I watched Percy first on *Gardening Club* and then on its successor *Gardeners' World,* little dreaming that one day I would follow in his footsteps, in spite of telling a friend across the street that I wanted to be Percy Thrower when I grew up. (I was conscious of having aspirations above my station.) Like me, Percy started work as a gardener on leaving school, first at Horwood House in Buckinghamshire and then at Windsor Castle, before eventually becoming Parks Superintendent at Shrewsbury.

He began broadcasting about gardening back in 1946. I first encountered him in my job as a gardening books editor at Hamlyn, working with him on such titles as *My Lifetime of Gardening.* (I remember being amused by the fact that in the bathroom at his house 'The Magnolias', there was a 'soap on a rope' in the shape of a carrot.) He went on to run his own garden centre and to broadcast on the BBC until the corporation decided that his advertising ICI garden products was inappropriate for a man in charge of their gardening output. Consequently they dropped him in 1975. In those days rules were rules and the BBC had no truck with advertising.

His voice had an unmistakably mellifluous quality – due in part to his lifelong fondness for a pipe of Tom Long tobacco – and many of us grew up with him as our mentor either in print or on television. The first gardening book I bought was *Percy Thrower's Encyclopedia of Gardening*. I have it still, along with indelible memories of working with a gardening legend.

## Water butts

Oh, do get some. Put them in place now and they will fill up during the winter. Connect them to the down-pipes of your house, your shed and your greenhouse and make use of the rainwater they collect, which seems so much more conducive to plant growth than that stuff that comes out of a tap. Making use of the natural resources that are readily and freely available to us as gardeners is every bit as important as doing our bit for the conservation of insects and wild flowers, and with utility bills going through the roof, there is a degree of satisfaction in getting something for nothing or, in this case, for the price of a water butt.

# Wild flower of the month

## Sweet violet (*Viola odorata*)

The only one of our native violets that has a scent, the sweet violet will occasionally open its flowers as early as December in hedge bottoms and the woodland floor. Its large, heart-shaped leaves are a tad on the coarse side, but its tiny violet blue, pale blue, or occasionally white flowers that nestle among them are delicate of construction and robust of perfume; though, strangely, after that first sniff it may seem to fade – thanks to a chemical called ionine, which temporarily incapacitates the olfactory senses. It was used as a stewing herb in medieval times and as a cure for headaches and insomnia. The Ancient Greeks valued it in perfumery and a sniff of that glorious perfume in winter will still lift the spirits. It has surprising tenacity when it is growing in a spot it likes.

*Other local common names*: Common Violet, English Violet, Florist's Violet, Garden Violet, Wood Violet. (Nothing very rustic there . . .)

# Poinsettias

Of all the plants that appear in garden centres at Christmas, the poinsettia is the most spectacular and the most popular. Once available only in bright cerise pink, it now comes in various shades of pink and cream, mottled or plain, to suit your décor and taste. Whichever you plump for, a word of warning: buy your plants from a warm store or garden centre rather than a chilly pavement display. The poinsettia is native to Mexico, a country not known for its frosts, and freezing temperatures chill them to the extent that their leaves fall off – a few days after you've got them home. So . . . wrap them up well for the journey and when you get home, position them in good light but away from a heat source, which can dry them out too quickly.

Keep the compost evenly moist and they will last well into the New Year. Cut back by half in spring (keep the milky sap away from your eyes) and it will turn into a green foliage plant. Short days and long nights (at least 12 hours) encourage the coloured bracts to form (the flowers are the tiny blobs in the centre) and in the home, that state of affairs is interrupted by the turning on of lamps. So if you want your plant to colour up for Christmas, put it in a cool, dark cupboard every evening for two to three weeks at 6 p.m., starting in early November, and take it out the following morning. Two weeks of long, cool nights, but bright light during the day, will trigger the colouring-up process.

Yes, I know, it is easier to buy a new plant, but if you want to know how to make your old plant do its thing again, that's the technique.

# Things you can do . . .

— Order seeds for sowing next year.
— Dig over the veg patch and make any new beds and borders – but take it slowly – half an hour at a time to be kind on your back muscles.
— Prune roses.
— Bring winter-flowering bulbs such as narcissi and hyacinths indoors, but keep them in a cool, bright place so that they last well.
— Force rhubarb outdoors under a forcing pot stuffed with straw.
— Protect cauliflower curds by bending over the central leaves.
— Prune fruit trees.
— Check stored fruit and remove any that show signs of rotting.
— Buy your Christmas presents before 24 December.
— Cut off the mass of dead large-flowered clematis stems back to around 1m from the ground. You can cut back to growing shoots in February.

# Things not to do . . .

— Any of the above if you just want a little time off. The prime reason for having a garden is for it to play its part in making our planet a greener place, but it should also give you great solace and pleasure. Don't forget that. And we are finally at the end of the year.
   Happy Christmas!

# Acknowledgements

My thanks, as always, to my editor Rowena Webb, whose idea it was for me to write this almanac and who has continued to offer boundless encouragement. Also to my literary agent Luigi Bonomi who, when I am halfway through a book and wondering if it was at all wise to begin, will convince me that it was.

I have confessed that I am a dipper as well as a reader, and there are several books that have been useful for dipping into during the writing of this volume. Some have been used to check the facts of which I felt reasonably sure but, at the same time, I needed to make certain. Others have offered information that I could not hope to retain.

The most frequently consulted is a book of which I am enormously fond, because of its scholarship and the research that went into its compilation: Geoffrey Grigson's *The Englishman's Flora*. I would take it to a desert island, especially in the first edition, which I treasure. It is a book of considerable size; it feels good, the paper is of wonderful quality, and it has what publishers call a pleasing 'heft'. It, alone, is responsible for furnishing me with the local common names of wild flowers. No one who loves our native flora should be without it and, if a first edition is beyond your means, settle for a paperback. Mine fell apart with over-use before I lashed out on the original.

*Birds Britannica* by Mark Cocker and Richard Mabey, the *Larousse Field Guide to the Birds of Britain and Ireland by* John Gooders, and *Flora Britannica* by Richard Mabey have also been great helpmates.

Others I have consulted are noted within the pages of this almanac and it will become apparent that my taste in books is both catholic and comprehensive. I admit to having amassed, over the last fifty years, a library that runs to around six thousand volumes. I mean, how many books does one man need? Some are valued friends, others are interesting acquaintances. Like everyone, I press a button and have access to the internet, but it remains a spiritual and tactile pleasure to open the covers of a book and feel its cover and pages within my fingers, and then to lose myself and be seduced by its contents.

The line drawings and jacket illustrations are my own, and it will come as no surprise to discover that I have only one 'O' level (that's one GCSE to those under the age of fifty) in art. It seemed a shame not to make use of it. My art teacher at school, David Wildman, would no doubt smile indulgently and pick up my brush to offer a few well-placed strokes that would transform them into something rather more accomplished. They exist here without his improvements, but possessed of what I hope is perceived as a certain rustic charm.

# Index

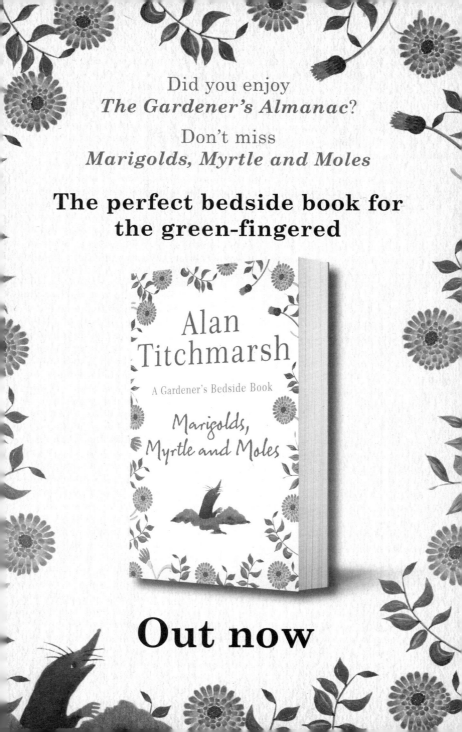

### An invitation from the publisher

Join us at www.hodder.co.uk, or follow us
on Twitter @hodderbooks to be a part of
our community of people who love the very
best in books and reading.

Whether you want to discover more about a book
or an author, watch trailers and interviews, have the
chance to win early limited editions, or simply browse
our expert readers' selection of the very best books,
we think you'll find what you're looking for.

And if you don't, that's the place to tell us what's missing.

**We love what we do, and we'd love you to be a part of it.**

www.hodder.co.uk

@hodderbooks

HodderBooks

HodderBooks